The British Birding Year

Two decades of written sketches celebrating the avian and natural spectacle of Britain. The birdwatcher travels the length of the country to record birds coming and going as the seasons change. Occasional rarities also make an appearance. The weather changes too: from the heavy snow of 2010 to the damaging floods of 2013/14, a wet and stormy trend contrasts with earlier, quieter years.

Andy Gibb is not *the* Andy Gibb but he did have the name first so he's sticking to it. Writing since 2004 and a birder since 1995, Andy first came to Bristol as a student in 1974. Yup, he's ancient as Hell and has been around a bit – Italy, Arizona, California, Australia and New Zealand. And what's it taught him? Not a lot but he has seen 1,258 bird species.

Also by Andy Gibb

Fiction:
Let the Time Come
Tales Told by an Idiot
Clearing the Brushes

Birding:
The Honeyeaters' Tree
Birding Peru 2015
Build a British Bird List

THE BRITISH
BIRDING YEAR

Andy Gibb

First published 2010

Fifth edition © 2022 Andy Gibb

ISBN: 978-1-4457-7242-4

To a Honey Buzzard Shot in Malta
(Maybe One of Ours)

Table of Contents

January

1st-19th, 2010 I kept scraping the car off and digging the drive out and then waking up to more. The snow, however, had been pushing birds in to the feeder over Christmas so there was an upside.

I'd also scattered food around because the ground foragers were probably having the hardest time. Nuthatches and coal tits helped them by turfing out seeds that didn't quite come up to scratch. They did a good job.

The persistent falls were enough to make it like a proper winter, which ended up being the coldest in the UK for 31 years. Its continuation into the first day of New Year kept me indoors until the afternoon.

I did venture out before daylight failed completely, which around Peebles is precious early, and I braved ice down to the Tweed. Fresh layers of snow were keeping pavements quite slip-free and walkable. There was a good crunchy, grippable surface to them.

So, an easy descent for me and paydirt for the year list with a couple of goosanders and an unexpected little grebe. It got better as I strolled further into town. A pair of bobbing shapes by the river's edge disclosed themselves as dippers – my first for the Borders and an excellent New Year addition. That short walk took the list up to 26 and all within a white mile of the house.

Next morning I'd hardly stepped a few paces back into town when the unmistakable profile of a peregrine falcon glided over the wee housing estate. Something of a surprise but perhaps Peebles has a resident bird. The town's iconic church may be the perfect home for it.

I continued my journey past the perfect homes of the Craigerne development, where I soon encountered the highlight of the year and a species I'd not seen since 2007. Goldfinches and siskins buzzed and tinkled to alert me to a mixed finch flock, which also held chaffinches, greenfinches and bramblings.

Yes, definitely bramblings with their orange breast band that wraps round to their wing coverts. They look as though they're ready for dinner with an orange napkin. Dinner being hard to find in the freeze probably brought them into our realm. Yet another cheer for snow in that respect.

I was trapped in Peebles by snow until a brief foray to Joppa in Edinburgh yielded velvet scoter, long-tailed duck and purple sandpiper. The roads still didn't seem safe enough to commit to a trip and it took another five days before heading north, via Edinburgh again, which added scaup, Slavonian grebe and knot.

Farther on were snow buntings at the Cairngorm Ski Area. A small flock is guaranteed from the car park in winter and they didn't disappoint. My usual motivation for driving up to such a commercial spot is easier access to the summit at Cairn Gorm and ptarmigans on the way. My usual date for this is in spring when the snow has cleared enough. January seldom cuts it and this one certainly didn't with deep drifts all around.

Now here's a thing: a snow bunting isn't a bunting; it's a longspur, like a Lapland longspur (or bunting as we call it). This is largely a North American family. Our other buntings belong to the Emberizidae, which also includes the numerous New World sparrows. Clear as mud, eh?

My ensuing route then also granted me black grouse in the Highlands and the annual snow goose at the head of Loch Craignish, south of Oban.

I dipped horribly on that town's regular ring-billed gull. I swear I saw it but didn't register it. However the harbour there is always good for the northern speciality of black guillemot, and hooded crow is certain somewhere round the foreshore. Shags and red-breasted mergansers were present besides, although both also occur throughout Britain.

10th, 2015 Last year was the warmest on record, not just for the UK, but worldwide – such a contrast to the winter of 2009/10 and the end of 2010. Indeed the four years between saw a mere two months clearly below the recent historic average and a staggering 29 a good way above. By recent I mean 1961-1990 – about 40 years ago, and in itself hotter than pre-industrial times.

This year has started warm and energetic. Gales have been swept in from the Atlantic by a jet stream that's misbehaving again. Violent swings in the weather have followed each time as though the status quo were breaking down and switching to some other equilibrium. This may not be conducive to comfort and we'll be in for a wild ride.

It was too wild at Cley Hill, between Frome in Somerset and Warminster in Wiltshire. This is a barren and exposed National Trust outcrop, so the wind whipped across it without mercy. I might have got on to a flock of linnets but I couldn't hold the bins steady enough to tell even such an easy bird. This was the signal for a retreat a mile or so south into Longleat Forest and relative calm.

It was also calm in the number of species. One upside was a walk well away from the rat-race in one of Wiltshire's rare woodlands, surrounded by pines, magnificent oaks and, as expected from the map, a lake.

Shearwater Lake, I discovered, and the name rang a bell. Checking back through my records unveiled a 1999 visit there and again only a handful of species. God knows why I'd logged it. At least the birds this year had some class: a little egret perched in a tree; crossbills were a first for the county, siskins called and a treecreeper was a welcome early addition to the year list.

One shore of the lake got the full benefit of the, by now forgotten, wind racing over cold water. I battled along it to reach a tearoom, which was closed. Ho-hum, back into the teeth of the gale and then the shelter of trees. I think we'll need them for all sorts of reasons in years to come.

20th, 2011 Amazing that it took a year in North Somerset's boom town of Portishead to see a goldcrest at the Port Marine development (more modern housing; where will it end?) but one flitted through saplings along Phoenix Way this morning.

It was close enough to show its golden crown stripe without needing binoculars. A gorgeous bird, and our smallest, despite the original Trivial Pursuit having the wren as the answer to that question.

Under the IOU (International Ornithologists' Union) taxonomy, goldcrest precedes the wren family and follows the Australian white-eyes. This makes sense for those who have seen all three. Yet this little bird's flash of colour sets it magically apart. Perhaps magic is how it survives our winter, especially the weather we had in the last month of 2010. It was even the coldest December since records began in 1910.

Also new yesterday were pochards in Bristol's Floating Harbour. Later were singing blackbirds and song thrushes. Spring must come early in the big city. Actually a song thrush was in full cry through nearby Eastwood this afternoon, as well as many great tits.

They might have been in for a shock in the ensuing weeks but that was it for cold weather. Indeed the 2011/12 winter reverted to warm, and dry, which was not without consequences.

20th-21st, 2010 Even more exhilarating for being entirely unexpected, tens of thousands of starlings were my reward for pushing on from Caerlaverock to Carlisle. Above the M74 they swirled through the sky, creating and undoing pointillist patterns on a whim. Gretna Green was the unlikely location, but one as renowned for the spectacle up north as the more southerly Somerset Levels.

Caerlaverock itself had been quiet although the

regular flocks of barnacle geese had dotted parts of the reserve. Buntings too had been notable with yellowhammers and reed buntings flitting through hedges near the visitor centre.

The day after Gretna continued the hard-to-get species south of the border. A bitterly cold few minutes at Heysham netted me twite and a Mediterranean gull, and finally Lancashire delivered, as usual, ruff and tree sparrows at Martin Mere.

A great start to 2010.

22nd-24th, 2016 Solid rain for an hour out of Bristol but the forecast had that travelling east faster than me, so I was hopeful that by Pitsford Reservoir...

Sure enough, Northampton was down to a few drops and a red kite on the road in to Pitsford bode well for my target bird. Which didn't manifest itself, or themselves. Male and female smew had been reported on BirdGuides, close to the causeway too. I braved a bitter wind off the water for half an hour but found only goldeneyes and little grebes.

No matter. My next stop, just into Norfolk was even less likely to deliver. A pallid harrier had been patrolling a bleak farmscape at Flitcham for over a month, having relocated from a previous month's sojourn at Snettisham. This was a bird that wasn't going anywhere in a hurry but even so, my record for the rarer harriers is zilch.

It stayed that way as a quick foray netted me red-legged partridges and a flock of tree sparrows, which were a first for the county. It was only as I drove on that I spied more cars and other birders. Damn! I'd been at the wrong spot but they too had drawn a blank. I'd try again on Sunday's return journey.

So to Saturday when harriers and owls topped the bill. My target was Hickling Broad's roost of wintering raptors. This National Nature Reserve lies a good 20 miles southeast of Sheringham where I'd roosted but after my rainy six-hour drive across from Clevedon, there was no way I'd make the morning show. So I had the day to kill exploring the coast down to Winterton.

Despite the route's lack of parking, I managed to see red-throated divers, common scoters, Brent geese and one gannet. Then Horsey Mere really started the ball rolling. It's one of the lesser bodies of water that constitute the Broad but it's still cloaked in reedbeds and a marsh harrier started my raptor list. The next species was so unexpected that it was the highlight of the trip: I hadn't seen one for a few years.

And I'd considered abandoning the circuit I was walking to the north of the Mere, so muddy was it. However I slipped and slid my way round in sight of a rough pasture that looked promising. And it delivered: just after midday in fine winter sunshine a barn owl quartered the scrub. It was one of those "Ye-e-ess!" moments and I took plenty of time to get the bird in my binoculars before it landed on a water trough. That called for scope views and I was glad I'd elected to lug the thing with me.

Almost as a distraction a kingfisher also perched in the foreground. That was new for Norfolk and the rest of the walk could only be an anticlimax. If you can call little egret, bearded tit (albeit heard-only) and Dartford warbler a bit boring. Actually the warbler was a county tick too. I finally tried to pick a reported bean goose out of a flock of some 2,000 pink-footed geese but the task was beyond me. And anyway it was time to drive halfway round the Broad to the Raptor Viewpoint, which was about a mile away as the goose flies.

This is signposted from the nature reserve car park but there should also be signs about donning wellies. Maybe this bit of Norfolk is so watery, it goes without saying. It's almost like the Broads are slowly reclaiming the surrounding land. Anyway, at the first puddle that spanned the entire lane I couldn't be arsed to turn back and change and ploughed a furrow down the middle of the road, being higher than the edges.

I was concentrating on this procedure when a shout of "Barn owl!" stopped me dead. A woman behind me was pointing and I followed to find a ghost pale bird threading its way into a nearby copse. No doubt then about my second sighting of the day. Just like buses...

The viewpoint was busy. Twenty or so cranes, countless marsh harriers, one ringtail, a stonechat, skeins of pink-footed geese and a short-eared owl so pale that the initial shout-out was for yet another barn owl. Even a bird as mundane as a redpoll was notable for being new to my Norfolk list. That made it five thus far and Sunday was to add another five. It was like I'd never been to the county.

Two of these came at the old stomping ground of Cley Marshes, which had gained a super modern visitor centre and extra acreage since my previous visit in 2004. A bittern loping over the reedbeds was the first rather tardy addition but a buzzard as I packed up to leave... Well, species number 156? For the premier birding county in the UK?

Then it was back to the "bleak farmscape" at Flitcham for another shot at the pallid harrier. I arrived just as folk were going, "There! There! It's just flying over the hedge!"

Yeah, but which freaking hedge? The hedge was doubtless famous to the assembled but to the newly arrived it was one of many and that was it for the bird's appearance. In compensation though a merlin sported on the distant horizon, fieldfares mixed with redwing flocks and bramblings with chaffinches.

So, this farmscape wasn't so bleak; check out Flitcham if you're passing in the winter.

24th, 2006 A red-necked grebe was twitchable at Duddingston Loch, practically in the centre of Edinburgh. The hardest to find of our Podicipedidae family, it also prefers the coast in winter so inland sightings are sparse. Indeed this bird stuck to the middle of the loch, as far away from shore as possible.

Rather more common, fulmars were already back prospecting nest sites on neighbouring Salisbury Crags. These overlook Holyrood Park, which I crossed on my way to the grebe. One hooded crow strutted his stuff there among the regular carrion crows. Hoodies tend to be the more western species in Scotland and are absent from England.

The previous week I'd found three dippers at

Colinton and the beginning of the month had delivered a kingfisher, unusually on Joppa seafront. Cold weather and iced-up streams can drive that open-water specialist to the coast but the start of this year had been warm for Scotland. As had December so who knows what the kingfisher was doing there.

2012 The easiest of the Podicipedidae – great crested grebe – brought up my hundred for Portbury Wharf Nature Reserve, on the Bristol side of Portishead. It's a late addition, for the sort of bird that loves that habitat but better late...

A mile further down the Severn Estuary, round Portishead Pier and the beach below the Royal Hotel, black redstarts had been reported during the month. A brief glimpse of one of the males was my first sighting since the same time in 2011. The species is a bit of a town speciality with five of my seven UK records.

Another mile on at Battery Point, three purple sandpipers made an appearance, despite continual disturbance there.

Farther afield, I managed to add the great northern diver at Cheddar Reservoir to my Somerset list before its demise in a tangle of fishing line. Then, also new for the county and indeed England being as it's an American species, was an obliging ring-billed gull near Woodford Lodge at Chew Valley Lake.

Even farther, Upton Warren, just south of Bromsgrove in Worcestershire, played host to flying snipe. An unsuccessful search along the River Salwarpe there for a reported lesser spotted woodpecker did turn up a kingfisher. On the way up the M5, Slimbridge had provided the usual Bewick's swans, white-fronted geese and golden plovers but its now annual lesser scaup was elusive.

At the beginning of the month I did something that has never occurred to me in nearly ten years of scope ownership. I turned it on the heavens and a unique clear night revealed Jupiter's four moons and equatorial bands right from my backyard in Port Marine. Definitely one of the 1,001 sights to see before you die.

February

1st, 2014 Another drive through copious rain, to get to Bedford this time. Why Bedford? Certainly not for its charms, as depressing a town as England offers. The centre hosted more police presence than punters out enjoying a Friday night. Wetherspoon provided the only civilised drinking and all the curry houses occupied one tiny little ghetto. However, the Magna Tandoori, which I finally settled on, did a nagalicious naga. So that was OK in the end.

No, Maulden Wood, to the south, was the draw. This is the last clawhold for Lady Amherst's pheasant – a bit of a trash bird for the British list but still fair game. I wasn't too invested in the result of searching for one and just as well since the wood, beyond the gravelled rides, was swamped and unworkable.

A chilly morning sun did bring out nuthatches, jays, coal tits and a splendid great spotted woodpecker. And a green woodpecker with its black eyepatch. One common pheasant also called from a thicket.

I then checked out nearby Stewartby Lake for a reported great northern diver. My route took me through the village itself. A-ha! I thought as I passed, this looks like a model village and I wasn't wrong. Built for the London Brick Company in 1926, it still stands though the works itself closed recently. The chimneys must stay all the same: they're listed. A pity we take more care of our industrial heritage than our natural.

A circuit of the lake proved once again infeasible and for the same reason. Its sludgy margins threatened at every step to slide me into the mire. I retreated to a McDonald's in Milton Keynes to check BirdGuides. I don't have to tell you how foul MK is, do I? Built on the American model that James Howard Kunstler so hates, it too will collapse with the disappearance of the cheap energy that underpins it.

Anyway, the news was of a cattle egret over in Calvert, Buckinghamshire. I sped away from the urban sore – about the only benefit of those highways criss-crossing it. Clouds were building as I reached the little BBOWT reserve so I hurried in to be greeted by...

More mud. I was definitely hitting the part of England that's seen record January rainfall. In fact from Calvert through to Swindon almost every field had its pool of standing water. Those bordering the A34 round Oxford had merged to produce a landscape that was more seascape. It was surreal.

Did I see the egret? Nah, but a flyover red kite was cool.

6th, 2010 The chaffinch has one of my favourite songs: it's so cheerful. It also marks the beginning of spring for me, so today's male was trying it on a bit, especially because it was very much colder than yesterday.

The chaffinch's scientific name is *Fringilla coelebs*, which means bachelor finch. This came about from the tendency of females to migrate further than males, who then dominate the northern part of the species' winter range. So, when old Linnaeus started this whole taxonomy thang and needed some cool sounding Latin words for it, he used this behaviour to describe the chaffinch. Linnaeus, being in Sweden at the time, would have seen all these wintering "bachelors".

I heard my bachelor on the Mariner's Path, which is the first leg of the Severn Estuary walk from Portishead to Clevedon. It dissects the defunct golf course and Sugar Loaf Beach – so named because it has no sugar, no loaves and God forbid you should want to sunbathe there. Curious, eh?

My late winter stint for the BTO Bird Atlas took me along the path, which meant I got to count birds instead of just ticking them. It's a full-time occupation and the darn creatures don't keep still, so the job entails some hefty guesswork. Never mind: soon we'll have so few birds they'll be easy to tally.

At least the chaffinch doesn't yet qualify for that prognosis, being on Birdlife International's Green List, which means it's not on the Red or Amber Lists. Red stands for declining by more than 50% and Amber for 25%. There are 52 species on the former; 126 on the latter; 178 in total. Which leaves a somewhat arbitrary Green List of 68 – something like 28% of British species but certainly a minority.

That's the state of the world, my friends. Something that only nature observers believe. And it's why birding, for example, is important. We're the guys watching the altimeter as the jet plummets to the ground. We could be excused for being a miserable bunch.

12th, 2011 Another shorebird has made it over to the marina side of Portishead Pill. This morning's dunlin could have been a Port Marine tick for that reason. However, the county (sorry, Unitary Authority) of Bristol actually covers every inch of the mud banks round the Pill, right up to the high tide mark. You'll hear more about this annexation later.

Anyway, one of these little sandpipers was picking around among its larger redshank cousins. It might have been a bird from the flock that feeds off Portbury Wharf. That's easy to see, even at the distance of the docks at the mouth of the Avon, as light twinkles off the birds' twisting and turning flight.

Spring is definitely winding up. A skylark was singing over the Wharf's spartina and one of our rock pipits did his parachute thing by the lock gates. I then climbed up to Eastwood, which rises between the marina and the Severn Estuary. A red admiral floated by and settled to absorb what it could of the sun. February seems awful early for a butterfly.

15th, 2004 BirdGuides had been reporting scads of rarities on the south Fife coast, so I went for them. The priority was a lifer, which seemed to have been at Leven for ever – king eider. It was easy enough to find, easier for being an immature bird. It was clearly a different beast from its attendant common eiders.

Also in the bay were velvet scoters, long-tailed ducks and red-breasted mergansers. A few turnstones and knots loafed on an outcrop, and rock pipits flitted by from time to time.

Next, it was east a few miles to Lower Largo to scope more of the bay. This produced a black-throated diver, Slavonian grebes, common scoters and (thrill of the day) red-necked grebes. I'd only ever seen one of these.

East again, Ruddon's Point was a tad disappointing – the same roster as above plus a couple of grey plovers. I also discerned distant razorbills and two stonechats obliged in the coastal scrub.

Finally, I hopped another couple of miles to Kilconquhar Loch – not an easy place to spell, or pronounce. Neither the reported smew nor red-crested pochard was on display. Plenty of common pochard and a few scaup made up for the absentees. Near the car park, a lone tree sparrow consorted with local house sparrows.

The day's final tally was 60 species – remarkable for a Scottish February.

18th, 2004 A beautiful, crisp day in the foothills of the Cairngorms. There wasn't much bird activity but the Glens are like that. I wonder if they were better when they had trees.

Anyway, the end of the tarmac up Glen Isla does provide a few Scots pines and a small plantation of other conifers. These held most of the day's birds, the most notable being one brambling in amongst numerous chaffinches. It's always nice to see mistle thrushes too and even better to hear them sing – a rather more melancholy tune than a blackbird's, I think.

A few ravens cronked overhead and bothered the

lone buzzard. No eagles, of course. And no red grouse this time. However, that other gamekeeper's favourite, the magpie, did surprisingly appear – only my second sighting for Angus.

22nd, 2014 The oddest thing about this year's inundation of RSPB Greylake is that, restricted to the small car park, I picked up nearly 30 species in half an hour. Two previous visits to the reserve hadn't yielded half as many, although 2012 had provided spotted crake.

Can we infer that floods are good for birds? Buzzards were plentiful and I guess they'll have plenty of carrion when the waters recede. A female marsh harrier also flapped her languid way over drenched fields. Teal, wigeon, little egrets, swans, herons, coots and mallards were predictable waterbirds.

Passerines also abounded, maybe attracted by a wee feeding station. Redwings, fieldfares and the year's first reed buntings were among the usual suspects. A blackcap also chacked away, invisible in a stunted tree, and as I left, a Cetti's warbler fired up.

True, my earlier visits had been in June and September and one may expect more of February at a wetland site. But it was very wet. I drove on to Burrowbridge and the end of the A361 under these conditions – a contrast from three summers previously. I hoped to go up the Mump to survey the waterscape but the car park was packed. Instead I got my view later from the A39 over the Polden Hills. I'd have to say that the floods' extent looked similar to 2013.

So why the fuss this time? Let's face it: the

Somerset Levels flood; residents should deal with it or get out. Why should taxpayers' money go to ruining the area's character with concrete defences and 24/7 dredging machinery?

In contrast the area round Shapwick Heath, farther north, wasn't as aquatic as on Christmas Day. And the birds were still good: two kingfishers, a great egret, the year's first pintail and gadwall and one hopeful bittern already booming. At dusk a massive flock of some 100,000 starlings crossed in the distance over Ham Wall but Lord knows where they ended up. Many observers were out for the sight but none tracked down the roost.

24th, 2011 The last toehold in the old county of Avon for nightingale and willow tit, Lower Woods is now at the north-eastern edge of South Gloucestershire. There and at nearby Inglestone Common in the summer, three pairs of nightingales cling on. They even manage the occasional successful breeding. Willow tits fare worse and many recent years draw a complete blank.

I visited Wetmoor, as I called the entire area, in May 1999 and found neither species. I assumed that I was in the same place – just south of the road from Wickwar to Hawkesbury – when I pulled over this year into one of several parking spots.

A calling buzzard kick-started the morning and soon fieldfares, redwings, bullfinches and a nuthatch joined the list. The first of three marsh tits sneezed and wheezed in the distance. Then the day bogged down – literally. The area's not named Wetmoor for no reason. It was muddy and hard to work.

Andy Gibb

I gave it a couple of hours. My return route took me back up a track from Lower Woods Lodge, where parking would have put me more into the heart of the area. Given the time of year, there was never going to be a nightingale but there was no hint of a willow tit either.

I did have a decent count of 22 species before retreating to the Beaufort Arms in Hawkesbury Upton for a pint of Bob. That's an easy beer to find in the Bristol area but drinking it close to its Wickwar brewery makes it even more quaffable.

March

1st, 2011 Worcestershire sightings of bramblings are rare, so make a note of Lodge Hill Farm, which is a mile or so along an old railway track out of Bewdley. I walked it and thought what a lovely train ride it would have made, occasionally high above Dowles Brook and surrounded by trees. We English don't experience much in the way of arboreal splendour but the Wyre Forest, which straddles the county border with Shropshire, is an exception.

I was visiting in response to reports of hawfinches. They were not to be but the bramblings at the Farm compensated, along with siskins and redpolls – those other woodland finches – and yellowhammers, unusual for the habitat. A pair of ravens also cronked over and a few small flocks of stock doves skittered above the tree tops.

No visit to the Forest is complete without trying for dippers so, farther on, I dropped down to brook level. Instant result: one bird that was even singing. Further upstream another dipper, hopefully female, was hunting.

This route committed me to a long circuit but it was priceless on a warming late winter day.

3rd, 2010 The highest tide of the year covered the green grass of Portbury Wharf. You'd think, with all such spring tides occurring when Sun and Moon line up, the very highest would occur when the Sun is closest to the Earth. This, at 147 million kilometres, happens in early January, but the big spring tides always fall a good couple of months later. In fact the rule is: after a full or new moon nearest an equinox.

This didn't explain how 2010's highest was so early because the full moon would be on the 15th, much closer to the equinox.

What was going on?

Several factors determine the tide's height. Yes, proximity of the Sun is one but its furthest distance is 152 million kilometres in July. A spot of mathematics translates this to an 8% decrease in its contribution to the pull.

In contrast the Moon, which drifts between 357,000 kilometres and 407,000, gives a 23% decrease in the Moon's pull on the oceans at apogee (that's space talk for being far away from Earth). It's a bigger puller, and the more so for being closer, by an extra multiplier of three in fact.

Could it be that the Moon is somehow nearer to us around March than at any other time? This is not amenable to casual calculation because the Sun, as well as the Earth, is shunting the Moon around. It's a complex three-body problem – an intricate dance that the Sun is choreographing between Earth and Moon.

It manifests itself in style in the Bristol Channel, which has the second highest tidal range in the world –

"What has all this to do with birds?" I hear you ask.

Go on, ask.

Well, the rising sea pushes all the waders dotted round the vast Severn Estuary nearer and nearer to the shore until that too disappears. Then where do they go?

Nowhere near me this morning, apart from a dozen or so turnstones, who clustered on Portishead's Battery Point. However, the high tide also pushed all the passerines off Woodhill Bay's salt marsh. The resulting mixture of meadow and rock pipits was good practice in separating the two. They're not so hard.

Not as hard as getting my head round the tides.

12th, 2012 What with my earliest singing blackcap two weeks ago; blackbirds yesterday; bees and red admirals in abundance at the end of February, including one that found its way into the house via my washing basket! With windows open to cool my room the last fortnight. With the prospect that this will be one of the warmest UK winters...

In retrospect that's how it turned out, by a full degree, and one of the driest in the south and east. A trip to Sussex later in the month reminded me of parched California and dosed me with unexpected sunburn! April saw drought restrictions imposed, in echoes of 1976.

Back to March, 2012: I went to what I'm calling Dowlais to try to beat my record for first arriving wheatear, which stood at the 21st. Dowlais Farm lies on the Severn Estuary coast south of Clevedon and it was chilly and misty and more seasonal than of late. My harbinger of spring didn't show (it did a week later) but skylarks and chaffinches were in song, which seemed a touch early.

Also during this month around Portishead, one stonechat reappeared at Portbury Wharf; the black redstart was still by the Pier; a male goosander dropped in for a day at the Marina; nine magpies sat on the roof of Gibb Towers. And I swear a female white wagtail was wintering on the High Street. She was a uniform pale grey from forehead to rump; not even a juvenile should wear that later than the autumn. The Web bears no reference to the phenomenon and I wonder if such birds are simply overlooked.

16th, 2006 The first puffins of the year arrived at the sandstone cliffs of Auchmithie, near Arbroath on Scotland's east coast. Eight of them bobbed around on the sea below their nesting burrows.

How complacent that paragraph sounds from the perspective of 2013. The count this year was zero, as it was up and down the coast in the face of an unexplained dieback. Sure, the scientific establishment explained it to the satisfaction of its paymasters. Starvation, severe storms, they said, as though these tough little characters couldn't weather life at sea. What they can't weather is our resources grab and pollution. How deep must one dig to uncover these roots? Not too deep, I suspect, but deeper than makes civilisation comfortable.

Farther north than Auchmithie in 2006, and blown a long way from its home in the USA, was a vagrant. One Bonaparte's gull had turned up at Ferryden, by Montrose. I had already made three visits and failed to locate the bird.

The morning of Thursday the 16th the weather

was bitter. I thought I was mad even to try again but as I sat shivering in the relative warmth of my car, out of nowhere a figure appeared, complete with scope and binoculars. I left my cocoon and had a word with him. He had caught the bus up from Penicuik to Edinburgh, then the train to Montrose and tramped the final couple of miles. Man, that's dedication.

I left after another quarter hour to warm up in Tesco and do the messages, and returned about an hour later. The stranger was huddled in what pathetic shelter a small boat could offer. Still no sight of the gull. I left him to it. I fully expected there to be a frozen corpse the next time I went back.

17th, 2010 Slightly downstream from Bristol's Avon Gorge, both a greenshank and three common sandpipers popped up on the North Somerset side of the river. Or did they?

The Bristol boundary lies along the high tide line and these birds were foraging in the mud, which must be below the highest of those lines. Worse, on closer examination the border actually runs a little inland. Let's not fuss about that. Keep it simple: Leigh Woods side, Somerset; Sea Mills, Bristol; anything down the middle, up for grabs by either.

In the event all the birds commuted between the two banks and so chipped in to both county lists.

The greenshank should have been on its way in from Africa, as should the common sandpipers (or maybe southern Europe). Yet these individuals had been around all winter. And the winter before... and...

This is certainly new behaviour for the common

sands. It suggests they're finding our coldest season more and more to their taste. Dare I say warmer?

I left the waders and walked back up Paradise Bottom. Seven-plus buzzards thermalled over and two ravens cronked by. It's a magical nook of Leigh Woods with its pond nestling under the guard of misty bare trees in March. Two months later bluebells erupt just before leaves fully clothe these sentinels.

Around mid-March lone migrants of several species have dripped through the south of the country – exceptional individuals who herald the shape of things to come. For most folk the first visible sign of spring about now are early wheatears returning from their African winter home.

Of the *Oenanthe* genus, they are close relatives of our local *Saxicola* stonechats – both in the IOU's Muscicapidae (or chat) family. In Europe this includes our familiar and loved resident the robin, which thus loses its traditional place in the thrushes. They do though immediately precede the chats so the association is still close.

The first wheatears may be on passage through to Greenland and even Canada – one of the longest distance migrations in the world. They cross 1,500 miles of ocean in one hop, staying on the wing for as much as 30 hours. For a bird not much bigger than a robin this is a gargantuan feat.

Along with them come sand martins, out of tropical West Africa to their breeding grounds in northern Europe. Then the earliest waterbirds arrive, largely to breed in England, as little ringed plovers seek out our freshwater and estuary margins.

At this time it's worth listening for a bird that's becoming increasingly resident. A blackcap may break into a brief warble; it's unlikely to be a visitor from north of the Sahara – yet. Indeed the bird may be preparing for its own migration away from us to Scandinavia or Russia. The chiffchaff too has begun to overwinter and may start its metronomic song from the middle of the month.

Now the traditional harbinger of summer enters the stage. The barn swallow wings its way in from Africa, the first one definitely not making a summer: our calendar is still at the end of winter! The last of the hirundines follows as house martins return to our eaves and cliffs although early individuals will be on their way to yet more northern climes.

The chiffchaff's close relative, the willow warbler, also spreads in from the tropics before the end of March. This small passerine's sweet descending song gives it away before the eye sees it.

Another bird that may sneak through without being noticed visually is the true white wagtail, a paler race of our endemic pied wagtail. These migrants winter in the Middle East and North Africa and are on their way to Scandinavia, Iceland and Greenland.

25th, 2016 Tell anyone in Bristol that you've been birding Staple Hill and even non-birders will look at you funny. Mind you, folk from Taunton or Chard will probably draw a blank on the name but Staple Hill is woodland between those two towns.

On the north scarp of the Blackdown Hills it commands views from the Quantocks to the Somerset Levels. It's also a decent patch to work as birds enter their breeding season. For a start, updraughts encourage buzzards to display and utter their plaintive mew. If I say four were thermalling overhead, that merely counted the nearby birds. My entire binocular field of view must have disclosed a dozen or more. One pair was also prospecting the local pine plantation, plummeting down to one branch in particular.

Passerines were also easy to see as they chased each other or otherwise advertised their presence. Thus, I logged my first treecreepers for the year, and wrens and a goldcrest were equally obvious. Siskins and my first 2016 nuthatches were vocal but hard to locate.

Farther east the A30 more or less follows the border of Somerset and Dorset. I've crisscrossed this region a good few times without once finding a decent birding spot. So, a more thorough search was in order and I drove a loop through Blackmore Vale to Shaftesbury.

With about the usual result. A kingfisher and grey wagtail by Sherborne Castle were my sole reward. The area qualifies for my "bleak farmscape" epithet. Oh, it all looks very English and picture postcard but farming has a nasty habit of imposing its patchwork quilt over any messiness that might support our native creatures. It's hard to equate all this apparent prettiness with an industrial landscape but that's what it is. Farming is an industry.

April

In the remaining wilder patches the start of this month should see ring ouzels heading for their rocky upland breeding grounds. Tree pipits will distinguish themselves from our resident meadow and rock pipits by parachuting from treetops with their chaffinch-like song.

Raptors get in on the act too: marsh harriers that have survived hazardous Mediterranean crossings rejoin the few individuals that choose the safer option of wintering with us.

The floodgates open. From mid-April we enjoy yellow wagtails, redstarts, sedge warblers and the second wader – whimbrels. Seabirds have been returning to their colonies although most have been visible offshore in ones and twos throughout winter. Not so another of the long-distance kings as common terns come from as far as South Africa.

Its cousin, the true champion, arctic tern is about a week behind among the later spring migrants to pour across our country.

1st-2nd, 2017 Was it an April Fool that sent me searching for spring migrants in a Lincolnshire woodland? God knows the habitat is hardly usual for such an industrial county but it should be all the more a magnet, no?

As it happened I was trying to work the fens to the southeast of Lincoln itself, much as one might in Norfolk or Somerset, but the only provision for nature seemed to be the odd scattered remnants of limewoods that once cloaked the area. Southrey is one such and that's where I started the day's birding.

The site did at least produce several firsts for my Lincolnshire list – coal tit, great spotted wood-pecker, bullfinch and blackcap. Chiffchaffs out-swamped the latter and together they they were the sole evidence of summer. Coppiced rides within the wood, primarily for butterflies, would surely attract other species as the month moved on.

That also ended the day's birding as my circuit round Bucknall, Woodhall Spa and Metheringham drew a blank. So the next day at Attenborough, just outside Nottingham, needed to be less of a fool's errand. I first chanced on the reserve in 2010 and thought it had promise and the intervening years have fulfilled that potential. It's become an essential visit despite the preponderance of humans and their canine friends.

The most egregious improvement was clear even in the car park, where Cetti's warblers chattered. Moreover they flitted openly enough for decent binocular views – a minute at a time, what's more! A complete character reversal for such perennial skulkers.

Again chiffchaffs and blackcaps were legion and my first sand martins of the spring flashed past, horribly backlit, but they couldn't have been anything else. The tail end of my visit rendered any doubt needless, as will be related.

Scrapes have also appeared at the site and one

redshank and one little ringed plover made use of the facility. Plenty of wintering ducks were still hanging on – pintails, shovelers, goldeneyes and goosanders. All the latter were redheads. Maybe the males leave earlier? On a feral note the Egyptian geese were pretty much tame but one red-crested pochard could have been overlooked, being asleep. She was so pale though that she was quite noticeable.

The only apparent downside on what became a perfect, warm day was the absence of the tree sparrows from seven years before. But they had just moved. The visitor centre now has a sand martin hide at the end of a short promontory and on this I heard the distinctive cheep of the sparrows. They weren't so easy to spot but a little patience paid dividends.

And the hide itself is well named. It's right next to a nesting wall and the birds zip in and out and hover and perch a mere yard from one's nose. In fact I feared for collisions with the window. Hopefully it's obvious enough not to be a danger.

8th, 2010 Ham Wall and Shapwick Heath reserves are pure magic. These two, plus Catcott Lows and Westhay Moor, form part of the Somerset Levels. It's a complex that's becoming as irresistible as Norfolk. And Norfolk is Mecca for birders.

The magic continued today with high-flying, courtship-chasing bitterns against a backdrop of hidden individuals booming in the reeds. Spring also manifested itself with displaying marsh harriers, scads of calling Cetti's warblers, early arriving whitethroats and my first sedge warblers of the year.

Add a strong supporting cast of little egret, water rails, black-tailed godwits, willow warblers, chiffchaffs, blackcaps and the long-staying great egret. Stir in the usual suspects. *Voilà! Magnifique.* And the weather was stupendous too. My face was tingling from the sun.

So, thank you, all the people whose hard work is restoring the Levels to their former glory.

11th, 2013 Is this a sign of things to come? After the coldest March on record, the rains returned to mark my first holiday from work for years. Those rains came on top of one of the wettest periods in England's history: as evidence, flooded fields on my way through Shropshire and Cheshire.

Matters would surely return to normal if I took a break in the Forest of Bowland and continued north the next day?

Not a bit of it. I was happily logging my sightings at the Hark to Bounty pub (also my night's lodgings) in Slaidburn when the power died. Not once, not twice, but thrice, and the third time terminally.

Silence descended. There's something the last couple of generations can't have experienced. I bet it drives them bonkers.

The weather and the power cut must be unrelated, mustn't they? I don't know. It's all beginning to seem a tad weird. Especially as I was musing earlier in the day at the birthplace of the Industrial Revolution (Ironbridge) on how that had coincided with nature's plummet from abundance to scarcity.

Still, a brief exploration of Slaidburn had turned up Lancashire's first red-legged partridge and lesser redpoll. That's modern day Lancashire. According to maps in the pub, West Yorkshire used to claim the area all the way down to Clitheroe. I suppose some battle or other saw a transfer of power.

Taking the moorland road past Stocks Reservoir the next morning had me and up into evidence of heavy snow and over to Yorkshire. There, an obliging red grouse posed just too far away from my camera. Where in Yorkshire was I? No idea really. I called it Lonsdale and hoped that was general enough to nail it.

16th, 2013 "Back to normal," said the checkout lady at Fort William Morrisons.

And it was too as wind drove sheets of rain in off Loch Linnhe. While England had been shivering in the unaccustomed March cold, the west of Scotland had been dry enough for wildfires. Admittedly wildfires largely caused by "landowners". They burn heather, some of them for grouse to breed enough victims for "sportsmen". Isn't the English language great for distortion?

Hmm, gone a bit off-topic here. The downpours had started overnight at the excellent Kings House Hotel on Rannoch Moor. I'd been in the bar so I hadn't minded, sneaking glances at German lesbians, so I'd minded even less. But that too is off-topic.

The hotel is at the gateway to Glencoe. And, boy, had that been alive with silvery filaments streaming down every hillside and a ferocious wind funnelling through gaps and cuttings. This battering continued north of Fort William, up the Great Glen, until the sun came out an hour later at Drumnadrochit. I was being tempted into Glen Affric and I duly went.

Of course the rain intensified to solid at the end of a half-hour drive along the glen. The journey in was surreal: drenched autumnal vegetation comprised brown bracken and leafless birches. A consequence of the dry spell, I suppose.

I sat in the downpour's pattering for a packet of Twiglets and a Mars Bar before admitting defeat. Mind you, beautiful though Glen Affric is, it's never delivered much in the way of birds. Three previous visits over fifteen years had yielded spotted flycatcher and tree pipit (unlikely mid-April), osprey and dipper. Those visits had also been free. There's now a £2 charge to park. A sign in the flexible English language explained why.

I headed on to Inverness but was bushed enough by Beauly to seek refuge there. Within an hour the sun was out and a brief stroll by the river (not recommended for the quantity of dog shit) brought red kite and probable in-flight pink-footed geese. They weren't a bad end to a frustrating day. Nor too a superb naga curry at the Shimla.

17th, 2014 I could be excused of fanciful thinking by visiting Theale, near Reading in Berkshire. How is that a worthwhile site? After driving lanes south of the village, I had to settle on the one parking spot I could find, by a canal picnic area. This adjoined scrub and the biggest lake in the complex – once dug for gravel.

OK, a willow warbler did show and, like everywhere at this time of year, blackcaps and chiffchaffs were bountiful. A couple of mistle thrushes flew up and a green woodpecker. Still, it wasn't looking worth 80 miles of the M4, when a jarring call caught my attention.

I listened.

Again. Unmistakable. A nightingale had to be just metres in front of me, in a tangle of brambles and bushes. It kept singing and I manoeuvred without success to catch a glimpse.

The song took on an eery harmonic quality. I'd not heard that before – very weird. It happened again. Wait...

There were two birds! Then a third started up from another direction. I was surrounded by nightingales.

In their jousting they began to show with cracking views of rather a plain bird but definitely a gorgeous rust brown tail. It was marvellous: Theale was actually fine; these birds had beaten my previous early sighting by eight days. The year list was moving on again; a sedge warbler and distant house martins kept it going.

19th, 2010 Terns should have been in: arctic, common, Sandwich or even black. That was the purpose of my walk to Battery Point, Portishead. Instead, on the way over, the mud banks at the Marina entrance held three common sandpipers and a lingering redshank. It's always worth scanning the canyons and rills that slope down from the spartina marsh to the Pill (Anglo-Saxon for creek or inlet): it's easy for birds to hide in tham thar hills.

That's what the terrain looks like through binoculars.

In addition, a garden warbler was warbling softly, not in a garden but further on through Eastwood. That little woodland really delivers. I didn't view the bird but there's not much to see: it lives up to its scientific name – *Sylvia borin*. But it does sing sweetly, like a decaffeinated sedge warbler, or a blackcap with the volume down and in extended play mode.

Finally, swallows and sand martins fluttered and swooped over the Boating Lake. It was all very summery, apart from a disgusting yellowish layer of murk over the Severn Estuary. One's immediate reaction is to blame Avonmouth but the wind has finally veered round to the west. Was it Llanwern, which still clings to some sort of steel-processing life? Something else Welsh? Cardiff? A volcano?

[Not a Welsh volcano obviously but Icelandic since Eyjafjallajökull had been blowing off through April and disrupting air travel.]

22nd-23rd, 2013 A wind has covered the whole country for weeks, from bitter easterlies in Somerset before I left for Scotland, through Highland gales, to Nairn and Largo Bay. All the time battling the blasts. Just standing up was exercise enough, let alone trying to walk against the more ferocious gusts.

The battering continued down the A1 out of Edinburgh, shoving my car around the road. I made Lindisfarne in one piece but just in time for the tide to creep over its causeway. It would retreat four hours later but my schedule didn't allow that so I dropped a little further down Northumberland to Bamburgh.

There, dunes just south of the castle afforded some protection, but at the cost of the occasional mouthful of sand. I fashioned a seat in the side of one dune to anchor the bins on my knees for a spot of seawatching (a scope was out of the question). Thus my Northumberland list registered its first long-tailed duck and common scoters. The latter were strung out in a flock of about fifty.

These are birds that should depart for the tundra soon to breed but those incoming for the same purpose have been scarce, probably due to the weather. A few swallows, chiffchaffs and the odd willow warbler through Scotland, and very early (for me) arctic terns at Nairn had been it. The outrageous news about the demise of puffins seemed also to be true: a brief stop at the reliable spot of Auchmithie revealed none.

Further down the northeast coast, Seahouses added more swallows and my first house martin of the year. I then ambled further south to... Amble. By then I'd had enough buffeting and booked into the Harbour Guest House.

We may have to get used to this wind business. If the atmosphere is heating up, it stands to reason that the weather will get a tad wild. Wet and windy Britain always was; drenched and stormy looks set to be the new norm – the new, shifted, baseline for those who deny climate change.

The blowing had abated a touch by the middle of the following day and the B1257 over North Yorkshire Moors National Park. The route should have compensated for dipping on a little bunting earlier. It should also have compensated for that dip's dismal setting of Elba Park – some country park monstrosity being built south of Newcastle. At least it's what will count as countryside for the residents of yet more bland boxes going up under the guise of "homes" – the English language again. Rare birds will, however, turn up in these benighted places.

And then not show in the hour that I had at my disposal. So I was looking forward to what my road map marked as a scenic drive over the hills. As it was, one word sums up the experience – roadkill.

If you want to see dead birds, rabbits and, worst of all, hares, the western moors are perfect. You'd imagine a healthy population of buzzards to feed on this slaughter but all I saw was one kestrel.

A couple of timber production units (Forestry Commission) on the way through were no draw either and I was soon at overpriced Helmsley. Towns like that will get a nasty shock when the price of post-peak oil does the next of its launches into the stratosphere. It will also likely ground fighter jets, one of which nearly gave me a heart attack by roaring past at zero feet. I guess this must also be a characteristic of the B1257.

In future I'll stick to the raptor-friendly Wykeham Forest section of the North Yorkshire Moors.

* * * *

24th, 2010 Portbury Rhyne is a drainage ditch running from the Bristol side of Clevedon through the intervening Gordano Valley and out to Portishead Pill. This is the inlet just by the Marina and it drains into the Severn Estuary.

Pedestrians can cross this rhyne near the Gordano Gate pub. Here is a thin margin of reeds and I caught a bird's jittery song from within. Initially I believed it to be a sedge warbler: the melody was rich enough. But as I listened, it somehow didn't seem that manic and I revised my opinion to reed warbler. I needed to see the bird and after a couple of minutes it obliged and a plain brown jobbie clung to a stem.

The plainness clinched it although this individual did have an eye-ring that gave it a slight supercilium. More reed warbler evidence showed in its warm brown rump – not a phrase to be used anywhere other than in bird identification! *Acrocephalus scirpaceus* it is scientifically, the family slotting between the leaf warblers (like chiffchaff) and grasshopper warbler.

Reports of one of those have come in from Portbury Wharf this month so I strolled over there. No joy with it but plenty of joy with a few tree pipits, a secretive lesser whitethroat and thirteen wheatears.

The next few days should see swifts and the Wharf looks right for hobbies too. I'm a little surprised I haven't seen sedge warbler there and common tern looks a reasonable expectation as well.

25th, 2011 In the context of Bradnor Hill on the Welsh edge of Herefordshire you'd think our resident meadow pipit would be more likely. This National Trust property is basically a golf course, i.e. greens surrounded by scrub. This is archetypal meadow pipit habitat not much blessed with trees.

Even so, one of the birds I encountered had – to quote from my scribbled drawing – an eyebrow, unmarked upperparts and obvious pale wing bars with clear spotting above the median covert bar. A-ha! you exclaim, sounds like that summer visitor, a tree pipit. And I'd have to agree.

Sadly, the bird refused to sing or display or do anything that would have clinched it. A meadow pipit can look like that in some of the particulars but in all four?

So, I'd actually gone to the limits of England to look for a reported dotterel. You'll learn how this was dip number umpteen and several. But Bradnor is a fantastic spot with views down to Hay Bluff and Waun Fach in the Brecon Beacons. Skirrid is farther to the east and up north lie the Long Mynd and Clee.

I rounded the day off with another disappointing run over Hay Bluff. That road down to Llanthony passes the worst of British moorland, chomped almost to bare soil by grazing. The next road west, up from Forest Coal Pit looks like it should be more productive but it doesn't run all the way through. Funny how birding changes one's perspective of the landscape.

29th, 2010 A bonanza at Portland Bill, south of Weymouth, which held my first guillemots, razor-bills, kittiwakes, fulmars and gannets of the year. Add rock pipits galore and a couple of wheatears before the star of the show.

The wheatears were in view when a large wader took off behind them. Somehow that bird looked all whimbrel but I couldn't say how, apart from having watched hundreds of curlews over winter. And having spent a great deal of Tuesday dismissing them down the Severn Estuary. I knew a curlew. I didn't know a whimbrel so well. Out of pure caution, I pencilled the bird in as curlew.

Back in Weymouth swifts were obvious at RSPB Radipole, where I also saw three Cetti's warblers. Outstanding. Normally one hears dozens before any sighting of the elusive little chappies.

Presumably the beautiful drake hooded merganser I later picked out in one of the channels was plastic. In fact, the jury is still out on this one. Hoodie has only recently joined the British list thanks to an individual on North Uist in 2000. Yes, it takes that long to decide, so I'll not be holding my breath for the Radipole bird. The UK400 Club (aka Lee G R Evans) is not admitting it. Still, who he?

When I later accessed the Web, I checked Bird-Guides and lo, two whimbrels were reported at Portland Bill. The place really isn't curlew territory and it is a prime incoming whimbrel site. So I counted my bird.

May

5th, 2010 Swifts were in, so a certain handsome little falcon couldn't be far behind. Most years I pick hobby up before swifts but this spring I had to wait until today. One perched distantly by the River Arun in West Sussex. Some previous visits to nearby RSPB Pulborough Brooks had provided the entertainment of several hobbies at once, hawking insects above my head. The reserve is something of a hotspot for the species.

It's also a nightingale hotspot but not a single one called during this year's brief stop. I did find an unusual garden warbler. It sang in the open to give cracking views. Such a plain bird but almost homely in its simplicity.

I'd stopped earlier at South Downs Natural Burial Site, where Mum is. Many warblers there too, including my first wood warbler for the year. And a first for Hampshire.

5th, 2011 Always good to get to a proper bit of sea for boosting the year list. I was aiming for Dawlish Warren but it was raining. So I skedaddled down the coast to the town itself where a mere quid bought me parking right by the water. The first bird to hove into view was big, white and obvious – a gannet, nice and easy.

The next species was more difficult. A couple of ducks flew in the mid-distance. They should have been red-breasted mergansers; in the event they were shelducks. Still, a few fulmars made their way onto the list instead. I hoped for auks nesting on cliffs to the south but couldn't spot any. A nearby shag softened the dip. Then the highlight of the day appeared.

They'd probably been there all along but it was only when I retrieved my scope from the car that they came into view. Initially five, then another five, then dozens of common scoters. When was the last time I saw them that far south?

The answer is 2009 in Cornwall as you shall see.

7th, 2004 Auchintaple Loch, at the mouth of Glen Isla in Angus, produced two cuckoos and several willow warblers and swallows to prove that summer was reaching Scotland. I knew it was well under steam way down in southern England, where I had lately caught up with a couple of hobbies, two garganeys and even a swallow of the red-rumped variety. This had been flying round a most unprepossessing sort of lake in Lydney, Gloucestershire.

In between, at Leighton Moss, Lancashire, a bittern had boomed its acknowledgement of the advancing season and a spoonbill had made it feel quite Mediterranean.

Back to Auchintaple: a wigeon was the sole reminder of winter before it too headed north or east to find a mate. More incoming breeders included common gull, curlew, lapwing and oyster-catcher. A similar cast was present at nearby Backwater – or is that Blackwater? The maps can't agree – Reservoir a few days later. An osprey joined them from some hopefully secret and safe location.

9th-13th, 1998 I'd made a solid start to my biggest ever month with 40 species at Upton Warren. Devil's Spittleful, an interesting sandy heath by Kidderminster, had added 8 more before a return to Upton boosted the total to 60 in the first week.

However, Scotland unlocked the bulk of the species. Yup, I went on holiday. In Angus, Auchmithie alone was good for another 20, with great birds like eider, corn bunting, grey partridge, fulmar, guillemot and puffin. The last three also went on to my world life list. This had only stood at just over 300, inclusive of New England, so there was plenty of room for improvement.

Then five days on Orkney drove the month list wild. I had a wild time getting there too. Not for me a flight but the ferry out of Aberdeen, the theory being to bag seabirds on the way over. It was easy to add kittiwake, great skua and razorbill to both month and life lists.

However, both also shuddered to a halt as a fog socked in. Indeed so thick was it that the boat's route through Scapa Flow disclosed not a shred of land until Stromness was upon us.

The next day though was clear and lifers kept coming with shag, arctic tern and arctic skua. Then red grouse and hen harrier joined both lists on the 12th. I still have vivid memories of the latter quartering somewhere near Hobbister, such a thrill that it was.

The 13th saw gannet become my 110th species for the month and I still wasn't finished. A drive back through the Highlands, taking in Glen Affric,

Loch Ruthven and Abernethy provided more northern specialities (as well as rather odd first buzzard and greylag). Twite and tree pipit were also lifers.

This put me on 125 by mid-month and presumably I went back to work. Another shift at Upton and a weekend down to Hampshire had me at 132. A rather strange mopping-up finale brought jay, whitethroat and a cormorant (on the final day!)

13th, 2009 In the birding news: a trip of dotterels up in Nottinghamshire. I was due to drive to Scotland. The perfect route for me to take? It would, had my car been ready, which was turning into a saga in itself.

So, I couldn't get to them. This is why they're a bogey bird. Over the years I have tried for them up Glas Moel in Angus, in the Cairngorms, on Carnethy Hill, south-west of Edinburgh, and latterly on Bredon Hill. Could another year pass without adding them to my world list?

Two days later and finally ready to go, I heard of slightly closer dotterels at High Neb in Derbyshire. Finding the spot proved beyond me and outrageous weather banged the final nail, and I'd dipped on dotterel again.

I didn't dip on dipper one hour later at Whaley Bridge: three flitted around rocks under the A6 road bridge.

[2010 and 2011, as noted, also drew dotterel blanks; I didn't bother in 2012 or 2013. Will 2014 deliver?]

* * * *

18th-19th, 2013 I last went in search of cirl buntings in November 1999. I stayed at a B&B overlooking Slapton Ley, which gave me Dartford warbler and Slavonian grebe, but the buntings were M.I.A. later that day.

Not so 14 years on. So completely not so that, on getting out of the car, an unfamiliar trilling call greeted me. It could only be one thing and the task was then down to locating the bird. This too was easy: a wind-blown male was calling from the top of a bush and made my 294th British species.

The original plan for this weekend was to explore the Devon/Cornwall border. The lure of cirl bunting, sort of on the way, had been too irresistible so I'd overnighted in Torquay and made Prawle Point by ten o'clock the following morning. This was certainly early enough to catch my singing bird.

The whole stretch of coast was a delight. A couple of Manx shearwaters offshore were a first for the year among fulmars and other seabirds. The gorse and scrub held linnets, whitethroats and stonechats. The sun even threatened to come out.

Then it was off to Bolt Head, just south of Salcombe, for a reported Richards pipit. No sign of that but the year's first (and only) spotted flycatcher was gold dust. It was distant and field marks were impossible but its behaviour was unmistakable. A wheatear was also noteworthy. More than all that, the coast there too was equally beautiful with gorse, outcrops, bays, inlets, limestone and defiles. Defiles. I like a good defile.

So, then the question was: what's up the Devon/Cornwall border?

The Tamar valley for one thing but first my route skirted Dartmoor National Park – sorry, Dartmoor Rural Exploitation Zone. That seems to be the purpose of such a designation. Screw as many dollars as possible out of a scenic landscape, even if it means destroying its looks with quarries, reservoirs, car parks and all the paraphernalia of industrial civilisation.

Still, Burrator Reservoir at least could be a magnet for waterbirds. Not at this time of year although surrounding woodland was alive with passerines. The sweet song of willow warblers filled the air and a blackcap belted forth near where I parked the motor. The actual moors brooded over the scene and rain threatened. The weather forecast had advised clearer conditions in the north of the county and I wended thence.

My wending led me deeper into more of the uninspiring farmscape that we call picturesque. The odd belt of bright green trees was a relief but nowhere seemed to promise birds. Through Tavistock and Launceston up to Holsworthy Port – a Saxon term for trading place – the entire border with Cornwall was devoid of interest. Bodmin Moor, further west, would probably have provided more joy but was beyond my remit and I skedaddled back to Portishead.

19th, 2004 A day trip from Dundee to Killie-crankie, on the A9 north of Perth, completed the summer passerines for Scotland. The ice-scoured pass here holds a rare remnant of the country's ancient semi-natural woodland. It's a place to find songbirds that can't go elsewhere. So, the year list gained pied and spotted flycatchers, tree pipits, redstarts and heard-only wood warblers and cuckoos.

19th-20th, 2011 Holford nestles in the northern folds of the Quantock Hills and was my first stop on a two-day trip to West Somerset. From the car park one is straight into stunning deciduous woodland. That's the stuff that should clothe far more of Britain. One early, notable bird of the day, a calling garden warbler, probably agreed.

A climb led into the added bonus of gorse-clad hills – two habitats for the price of one, and such productive habitats too. I may have been a bit late in the day for hilltop regulars but my return to the area the next day fixed that.

On this beautiful Thursday morning I dropped back into the oaks and birches of Hodder's Combe for an immediate payoff. At first I dismissed a two-tone whistle as yet another great tit call but as I drew near, notes tagged on to the end put me on high alert. A song I'd never heard, which meant it could only be one species. A few moments with the bins brought a chirpy pied flycatcher into view for my first county record.

The Combe provided plenty of wood warblers, another deciduous specialist and again new for Somerset. It also furnished a calling cuckoo, which would become rather common by the end of the trip. In an attempt to locate it, or possibly another wood warbler, I happened to spy a tawny owl. It was my second sighting in the year, which was almost unprecedented. They must always be around; one hears them enough but one rarely sees them. And when one does, it's such a treat.

That was just the morning. I pushed on to the National Trust's Holnicote Estate, near Porlock, and worked Bossington Hill, overlooking the Bristol Channel. Wheatears were my reward.

Overnight at Porlock itself yielded my first Somerset dipper and a most excellent lamb and venison pie at the Castle Hotel. For a curry man that's a compliment indeed.

My first stop next day on Exmoor saw a yellow-hammer just beyond Webbers Post at Luccombe Hill, and then another back in the Quantocks at Crowcombe Park Gate on my last stop. Both sites and Simonsbath, so far to the west it was almost in Devon, carried the sound of distant cuckoos. The day though was being less than kind in the weather department and driving filled most of it.

Somewhere in the tour I passed a red-legged partridge, which also pushed up my county list. It's always an easy species to log from a motor.

25th-30th, 2012 A baby blackbird was a guest in our Portishead backyard. God knows how he (if male he was – I think so) got there: it's an enclosed death-trap typical of the Port Marine development. There's certainly no way out for a bird that can only flutter a few inches into the air. His absence of tail wasn't helping none.

However Mum and Dad were in attendance and the very jail-like attributes of the yard did serve to keep predators out. I mean cats of course.

By day three junior could flap up to the one piece of garden furniture – a brown sawn-off trunk – so he was well camouflaged. One tiny patch of dense greenery served as refuge during the night and in the heat of the day. The only modification I had to make to help him survive was to keep housemates out of the yard so that the parents could feed him undisturbed.

I was almost whisking him off to the wildlife rescue centre at Secret World on the first day though. I didn't see the adult birds for several hours and they also disappeared for similar periods during the youngster's stay. He spent a while refusing food too but this appears to be natural (if he's full, I suppose).

It was fingers crossed all the time that he would fledge fully and it was a shock on the last morning when he was absent. I scrabbled through the greenery in case he'd expired there and drew a blank and a sigh of relief. After that much attachment to his cause, I felt like a surrogate parent myself.

31st, 2009 I drove across the Cheviots from Peebles to Teesside for a juvenile purple heron at the new reserve of Saltholme Pools. This lies surrounded by the industrial complex north of Middlesbrough and is perfect according to the theory that birds will go where humans don't. Think of the army ranges on Salisbury Plain, the vicinity of Sizewell and Dungeness nuclear power stations and so on.

Well, almost perfect. A few of the local neds had decided to scare off the heron. Really. This is what their lives are about.

I wasn't too gutted because at close-of-play, as I was leaving the reserve, up popped a yellow wagtail, which I'd last seen nearly ten years previously. This was so far back that I had some trouble with the bird's identification. It's not uncommon, so it's hard to say how I'd managed to miss it all that time, apart from my lengthy spells in America and Scotland.

June

1st, 2010 No-one listens. It's that time of year when baby birds abound and look abandoned. "Leave 'em where they are or, if in immediate danger, move them nearby," is the answer to the perennial question of what to do with such creatures.

So, what did I see at Eastwood in Portishead today? A woman showed me a fledgling blue tit in her cupped hands. She'd found it "in the road." Poor mite, condemned to death when it could have had a fighting chance. Ignorance, not cruelty, will kill us all.

4th, 2010 British seabirds, for which the country is internationally important, breed in profusion at Bempton Cliffs on the coast of Yorkshire. In fact the spectacle is so grand that on no account should you miss it. Further specialities are yellowhammers in the surrounding farmland and tree sparrows on the RSPB Centre's feeders.

Another marvel of the area is a walk through typical English arable farmland from the railway station. This puts it within public transport reach of Scarborough, Bridlington, Hull and beyond.

I confess that I drove and then followed obvious footpaths from the Centre to the cliffs. The shortest of these jagged a little right to overlook a defile (that word again!) and the best spot for soaking up the sight of nearby puffins. Despite being busy, several such good viewing locations dot the cliffs. And at no disturbance to the birds.

I rounded off my visit at the rather excellent Ethical Catering van, which does all sorts of food, including great coffee. It was tempting to spend an extra day to catch the *Yorkshire Belle* on its cruise from Bridlington Harbour round Flamborough Head. It would have been more tempting had it been Saturday when the RSPB charter it to pass right under the cliffs.

But the raptor viewpoint at Wykeham Forest was calling with its lure of honey buzzards and goshawks. In the event only one of the latter showed in the distance. Well, I was past the peak time for them.

8th, 2012 Overnight gales continued into the morning and coincided for the first time with a trip up to Kidderminster. A chance to drop in to Severn Beach then and boost the old county of Avon list with seabirds.

Before that I gave Portishead another crack and sat in the car overlooking Woodhill Bay. This is not a journey I'd usually drive but the motor was going out anyway. A high tide was up and waves rolled over the marsh, farther than I've seen. Rain blasting in from the Severn Estuary reduced visibility but quarter of an hour did produce one gannet. That went on the Bristol list because it occupies the Estuary up to the low tide mark.

Then it was off over a speed-restricted Avon Bridge to the Beach and much better viewing but still not much activity. Some maybe Manx shearwaters flew up near the Second Severn Crossing then the birding improved with obvious fulmars passing closer – many of them. I say the viewing was better but the car was a-rockin' and a-rollin' in the high winds blowing from the west. My usual steady mobile hide was distinctly unsteady.

One small dark individual flying fast over the waves put me in mind of a petrel but I couldn't be sure. A flock of black shapes resting on the choppy waters had me reaching for the scope and thinking of scoters. They weren't. They were skuas, most likely arctic skuas with the square-ended tail I could discern on one. Pity I couldn't turn them into the poms that were reported about the same time.

The count was rising but still tentative when a clear storm petrel, with white rump like a house martin, chased upstream and more shearwaters wheeled in the distance. That made for a nice variety if not massive numbers.

The rain came and went and came again with only the addition of a distant gannet so I called it a day (about half an hour actually). It was but one soaking day in the wettest June on record – indeed the wettest spring if one takes that to run from April to June.

10th, 2010 Not much at Heron's Green, Chew Valley Lake: I drove on past the Blue Bowl and the Crown at West Harptree and round to the causeway at Herriott's Pool. No ducks there and hardly any gulls either.

That left Stratford Bay hide, in the permit-only section. Now, I'd been seeing many swifts from both previous vantage points but the numbers above this last location... At first I called them dozens. Then I had to go to hundreds after an attempted count. Then walking to the hide and back, I realised that the birds stretched not only from horizon to horizon but across the entire lake.

Thousands. And if I say two thousand, I believe I'm selling the spectacle short. I had to marvel: swifts filled the sky.

11th, 2006 An intrepid seven embarked on the *Maid of the Forth* in search of puffins and other denizens of the briny. In spite of assurances from he-who-should-know-better that they lay many leagues distant, several "parrots of the sea" bobbed round the Forth Bridge; indeed they were ever-present on a three-hour cruise in mill-pond conditions and a heat haze. Fulmars and gannets also seemed to escort our boat on its voyage.

Sandwich and common terns soon made them-selves obvious before Inchcolm Island (in Fife) introduced many to the fact that gulls are not just seagulls but five different species breeding in the Firth. Eiders and their cuddly crèche of ducklings provided an *aah* moment. Then a more sinister side of life on the ocean wave manifested itself in the guise of a great skua. This too likes baby ducks but more as an appetiser to its main course of pilfered fish. A couple of herring gulls policed the pirate away from their eggs, thereby providing a sort of community service for the residents of the island.

The passage from Inchcolm to Inchkeith (also in Fife) revealed guillemots and razorbills – from the puffins' auk family – and the first Manx shearwater gliding, albatross-stylee, a foot or so above the water. The kittiwake and shag colonies at Inchkeith added more species to the trip list as did the day's sole great black-backed gull. The journey back to Hawes via Inchmickery and past the distant Manhattan skyline of Leith recapped most of the birds hitherto seen.

15th, 2014 An unexpected bird is as good as a lifer. Especially when that bird is as classy as a short-eared owl. One certainly livened up a drizzly Axe Edge Moor, above Buxton in Derbyshire, at arguably the most southerly extent of the Pennines.

It's certainly a recommended spot for breeding golden plovers. Having missed them wintering down south, I was on their trail and had heard their plaintive piping but not yet connected with a sighting.

A large brown raptor flew up from the boggy grassland.

Buzzard! I thought and swung the binoculars up. (It's a good thing I still get excited by buzzards – legacy of being a twenty-year birder.) Anyway, not a buzzard... harrier? The bird was quartering but its colour was all wrong. With no ring tail or paleness but sporting massive wing patches like RAF roundels, only one candidate then remained.

I saw the owl twice before confirming that the piping did indeed emanate from golden plovers. One flew over my head; then another perched on a tussock right by the road.

Red grouse flushed from under my feet during the hour or so spent on the moor. They weren't a surprise, given that Axe Edge is a grouse moor, but I hadn't been anticipating them.

I motored on to spend the night in Wakefield, then headed farther north.

Into roadkill, just like the North Yorkshire Moors in 2013. It litters the routes over English uplands and maybe demonstrates how plentiful wildlife is up there. Or do drivers deliberately mow creatures down? It is possible to drive wildlife-rich terrain without hitting anything.

This year my more westerly route took me to Scar House Reservoir on the very eastern edge of the Yorkshire Dales. It lies under Great Whernside and Little Whernside, which are not to be confused with just plain old Whernside. That sits above Lonsdale on the other side of the Dales.

My original plan was to bird round Lofthouse but the draw of a road running into the Pennines to nowhere but a body of water was too much. I love these dead ends. This end, at the head of Nidderdale, wasn't so dead.

A couple of hours working the wee area round the dam netted 28 species. Among these was my first spotted flycatcher for 2014. June 16 is not an unusually late date for this late arriving species but it is very, very tardy for lesser redpoll.

I just hadn't seen one all year but a small ever-green plantation under the dam must be a breeding spot. Not that my redpoll was there; it was picking around in short grass nearby. I didn't have to strain upwards for a fine view of the dusky red forehead that gives the species its name. Normally that's hard to spot and one has to rely on the darker head and tiny bill for an identification.

Very, very, very late is June 17 for snipe. But the

next day brought my first year sighting, on Redmire Moor, north of Leyburn. It was bizarre. The bird was resting on a small wooden platform by the roadside, so it was easy from the car. Quite what purpose the platform, overgrown with grass, was serving is a mystery.

Mind you, that moors road to Grinton bisects an odd area, dedicated to tanks, firing ranges and other pursuits of war – doubtless getting a good workout preparatory to Iraq Mark II. Like the similar Salisbury Plain, this martial activity has one desirable consequence of excluding humans, so wildlife has a chance to thrive (unless it gets too close to the road).

War, what is good for? The natural world, it seems. Bring it on!

18th, 2014 Nine years had elapsed since my previous visit to Mull of Galloway – time enough to erase any memory of what this headland is like. A rather fine lighthouse and walled garden preside over a vista of deep blue seas and dark green grass. Between the latter, climb 80-metre high cliffs. The scene is about the last unspoilt section of the Galloway coast.

Seabirds abounded. Continuing the theme of surprises, I was unprepared for their numbers.

The cliffs do form a breeding colony, and guillemots, razorbills, fulmars, kittiwakes and shags predominated. They were all good grist for the year list mill. Manx shearwaters also glided farther out and a lone black guillemot fluttered past to remind me that I was on the west coast of Scotland.

The wind also reminded me, dragging tendrils of

a sea fret over the promontory. This bank of cloud obscured the Irish coast but the Isle of Man rose clear to the south.

An hour of buffeting sufficed, especially at the end of a day's drive from Dumfries via Loch Ken. I'd hoped the Threave osprey might venture up there or the few remaining willow tits would show. Neither obliged.

In general for the birder, this and next month can be the cruellest, breeding birds lying low, bringing up baby, singers mated and migrants migrated. It's a time of year when the BirdGuides map is blank, when butterflies become the new birds. Even rock pipits seem to desert their stronghold of Portishead.

Woods and hedgerows are silent. The odd furtive shape may flit from one branch to another but it keeps foliage between itself and the observer. The bird doesn't peep even a contact call to offer a clue.

It's a time when our estuaries and lakes hold only gulls, a few resident ducks and grebes, and ever-present coots and moorhens. A feral wood duck (the American one) in the centre of Birmingham (not the American one) becomes a highlight.

In July, birders' lists will hit a trough.

The first swell in the calm will arrive before the end of that month and on through August in the form of smart green sandpipers. But their passage will be a ripple: the real wave won't rise until mid-August when a few wood sandpipers depart their breeding territory in the marshy taiga.

They signal the start of the flood. Then

greenshanks roll in from their taiga nurseries in boggy grasslands and moors of northern Europe – a thousand or so even from Scotland. All these *Tringa* waders start the movement but other families, like terns, will be turning too – black terns first.

Somehow this stirring coincides with the biggest human event on the birding calendar at the end of August but we have a while to go before that – a while not without its unexpected crowd-pullers...

25th, 2009 "You timed it well," said one of the watchers strung along the path at RSPB Otmoor, just north of Oxford.

Alle-bloody-lujah. At last, a twitch where I turn up just as the bird, a marsh warbler this time, has started showing. It had been silent but now its song was clear and unmistakable. After fifteen years of trying to separate reed and sedge warblers by call, this one was a cinch. I still couldn't see the bird but a shape flitted in the depths of bushes.

Then a little brown job flew out and across the path to disappear in reeds opposite. It looked right for marsh warbler but it sped way too fast for positive identification. The same song began again in those reeds. So, unless there were two birds taking turns... no, let's not even go there. I saw it.

I saw it again later on my way back from scoffing lunch further along the path. The warbler made its reverse journey, almost as a curtain call to the performance whose overture I had seen.

The visit had started well with whitethroats *scritchy-scratch*ing their song from bushes and wires and a hobby powering by overhead. Later, I had four of these magnificent raptors in the air at once. They're the most elegant of falcons.

Also on the way in, both sedge and reed warblers had lined my route and on the way back a stonechat perched obligingly.

For the grand finale I had been told of a turtle dove around the car park. I listened out and caught its soft trill but couldn't nail the source. I searched and persistence paid off when I spied the individual, quite plainly out in the open at the top of a tree. It was much farther away than I had thought – always a tricky one, judging distance.

29th, 2016 Having spent eight solid months transplanting the software for a motorway sign from Windows to Linux, we'd reached acceptance tests at Gatwick. The client then had an unavoidable day off in the middle, so I took a rare opportunity for my own break and headed to Oare Marshes on the north Kent coast. This effectively saved me 300 miles of a 400-mile round trip (and the company paid for all the petrol!)

Of course it rained on the evening drive over but didn't flood as it had in the preceding month round the London area. Images of cars submerged in water were becoming all too common.

Rain was also in the forecast on the morning that I left Faversham for Oare but my biggest problem was finding a route there. A distance of about two miles clearly didn't demand signposts although the maze of one-way roads probably defeated even the natives. The rental car's Sat-Nav couldn't cope, so I was down to trial and error. Once in Oare it should have been easy, but no. Not a single indication of where the reserve might be.

A convenient Sainsbury's gave me the chance to ask a lady at the fag kiosk and she pointed me back the way I'd already been. It took just a few hundred yards more to get to the marshes. So, for future reference, aim for Harty Ferry and keep going until the road runs out. That's where the car park is.

A walk round the East Flood passed most of the best habitat and produced sedge and reed warblers, still vocal in late June, a whitethroat, an island full of black-tailed godwits, another of not so many avocets, one faintly pinging bearded tit and some early returning curlews. Little egrets outnumbered grey herons by about ten to one; how times have changed!

Preceding the rain, the wind was strong backing into fierce and made the scope pretty much useless. I dragged it farther into the West Flood where a marsh harrier was at least obvious enough for binoculars. Likewise a lone sand martin. I suppose you would call this quiet birding but it was fine and relaxing and an antidote to modern life.

The rest of the North Kent coast was too built-up for any extra birding action but Foreness Point, just beyond Margate, surprised me. A large area of rough grass was home to house sparrows, meadow pipits and what I first took to be a grasshopper warbler. The song though was too abbreviated. A moment's thought was necessary to get to corn bunting, which was still a major surprise for a sea-side town. The sound rang out again, from more than one spot, and confirmed my ID but getting a visual was difficult with so many sparrows and pipits flitting to and fro. Persistence paid in the end with a view of a male bunting atop one of the few shrubs in the area.

Information boards noted that fulmars bred on the cliffs below but I was frankly dubious, again from the built-up feel. However, they're too good to miss and despite the odd spatter of drizzle it looked worth continuing round the promontory to Botany Bay.

The species' flight is unmistakable at a distance and I didn't have to go far to see that, so another ten-minute walk put me in spitting range of several coasting at cliff-top level. The odd tern also passed offshore but never close enough to be certain they were Sandwich terns.

The rain continued to hold off as I returned to the car, found a cream tea at Ramsgate and then drove narrow country lanes to Stodmarsh as part of a detour back to Faversham. Here common terns flapping past the hide were certain and dozens of sand martins swooped betwixt reeds and pools. The site looked worth several hours instead of the few minutes I could spare before battling with the Canterbury traffic. That's one feature you can rely on in the southeast – congestion.

July

1st, 2009 My UK list includes a few birds I've not seen for a decade – a legacy of spending most of the century in California then Scotland. Heard-only nightjar was one such. I had thought of searching for the species on Cannock Chase but quailed at the size of the place.

Today, though, I caught wind of an RSPB evening visit, which seemed perfect for pointing me in the right direction. An amazing 30-ish people gathered at the Katyn Memorial and trooped off to Sherbrook Valley, as being the most likely spot. We were early, by at least an hour, but we also hoped to see woodcock.

The hour came and went. We moved back closer to the car park and, apart from one hobby, our staple fare was wood pigeons and crows.

Dusk fell. A couple of stonechats bush-hopped and a distant cuckoo fired up, rather late in the season, then shut down again. Still we waited for the nightjar's churring.

Nothing. Even after two more shifts of location nearer to the cars. It was ten o'clock. Some of the party called it a day; some of us tried a final time, across a road, past a café.

We had strung out by then and a handful of us got lucky: a couple of *pissit*ing woodcock sped past, like one huge bat in pursuit of another. It was over in an instant and a further half hour's wait brought nothing else.

I reflected that woodcock for a day hardly made it a waste. Doubly so because I had dropped by Gailey Reservoir late afternoon for a black-necked grebe, which did allow high-magnification telescope views. Like curlew sandpiper and red-necked phalarope earlier in the year, it was my first time for the species in breeding plumage – a bonus. And another species not seen in this country since 1999 – almost the lost decade.

We gave up on the nightjars and, almost at the point of no return, heard a faint call and scurried back. Finally, two birds were serenading. I didn't get to see them, again; like quail, they just seem to be that way. I wouldn't even have heard them if the camaraderie of the group hadn't kept me on the Chase for three hours.

People can be good for some things!

1st, 2011 Situation normal for a species that I'd only listed once, in 2005 near Winchester. It hides out in cornfields and issues its *wet-my-lips* calls and you'd have to yomp through crops and flush it to get any more. Which wouldn't please the farmer and wouldn't please the bird, so hearing is as much as one should hope for.

Actually, I lie. I heard plenty either of caged birds or tape lures in the benighted island of Malta, so the call of *Coturnix coturnix* also reminds me of that dismal Mediterranean lump. Nice to have some individuals escaping the bastards' guns to brighten our summer. So it was that Cissbury Ring, West Sussex held a calling quail.

The previous day on the way across Wiltshire I

quartered Salisbury Plain for likely stone-curlew sites. The best looked to be Porton Down but stopping for a squint would have branded me enemy of the state to judge by signs decorating the roadside. RSPB Normanton Down should have been more accessible but it's hidden better than the Army could manage. I found one possible path into it, walked a couple of miles, then gave up.

Guess what? I'd stopped within 200 yards of the reserve. A map would have been handy.

8th, 2010 Golden eagle would sound nice as the headline species of an ascent of an island's highest point but consider this: the 2007 Arran Bird Report notes only half a dozen sightings from thousands of contributions. These may not be all the submitted records and certainly they don't include the known nest sites. Quite right too: persecution in Scotland is an ongoing problem.

In any case, what chance a sighting in a few footbound hours, still kilometres away from the ridges and peaks that the eagles prefer?

Imagine this too: try to spot a human being on one of these ridges. They would have to be exactly profiled against the sky and even then would only show as a dot. An eagle would also be just such a dot, even with its wings in full stretch, albeit a dot in motion and probably detached from *terra firma*.

Ptarmigan would be nice too – just like that shown on an interpretive board near the start of the Goat Fell climb. Back to the 2007 Report again and just one sighting there makes it less than likely. Improbable, in fact, for a bird far more cryptic than an eagle.

Such is the stuff of hope but it's still a powerful motivation to visit Arran. Besides, Manx shearwaters and gannets enlivened the crossing from Ardrossan. Black guillemots bobbed in the harbours at each end, with breeding red-breasted mergansers at Brodick.

A host of woodland birds accompanied the first 300 metres of ascent. Then the moorland started and the day visitor can stop. At this stage, when the meadow pipits kick in, that's your lot all the way to the summit. It's barren habitat and, frankly, a bit boring birdwise.

Fortunately at this juncture a different path led back to Brodick Castle, past a wee hydro dam. Here were roving coal tit families, close views of a buzzard and, at the tiny pool, a grey wagtail.

No star species then but a fine haul of birds all the same.

10th, 2009 At least eight handsome green sandpiper chaps (or, more likely chappesses, but there is no visible difference) dotted the Flashes at Upton Warren, Worcestershire. In bird world it's autumn already and these individuals are returning from their nesting grounds, leaving their mates, generally the male, to raise the young. Women's Lib in full flight.

You see? Us blokes doing all the hard graft again. The distribution map in *Collins Bird Guide* shows the sandpiper's nearest breeding range as Norway or Sweden, maybe Finland too. So these ladies had flown over the North Sea to our freshwater margins on their journey back to Africa.

Also recently in from breeding closer to home,

on our own upland moors, were a dozen curlews. Several were identifiably young. Upton's own little ringed plovers seemed to have one chick – an improvement on last year, if so. Only one avocet remained and one returning shoveler in cryptic eclipse plumage nearly escaped detection among the mallards.

Far sadder was a lone starling. Have matters come to such a low ebb for starlings that they can no longer find huge gangs in which to roam? It seems almost impossible that these birds could become endangered.

A couple of colourful linnets completed the list of species outside those normally resident.

12th, 2010 Red kites are branching out from their Rhayader stronghold. The last time I was up the Ceredigion coast, albeit back in 1998 at Ynyshir, I had no record of them. Now you'd be hard put to miss the species. Ones and twos all the way down from Machynlleth to Aberystwyth, then inland to Devil's Bridge, where six were in view at one time. Over the Cambrian Mountains to the birds' highest density round Rhayader itself and on almost to the Brecon Beacons. Kites, kites, kites.

It was marvellous, just like being abroad, which is a sad indictment of the British attitude to raptors. In fact Wales is also like being abroad by reason of having trees – another sad indictment. From Conwy down to Betws-y-Coed onwards and even at the Severn end of the country between Abergavenny and Chepstow. Trees, trees, trees.

An island at RSPB Conwy had held fifteen little egrets and turned its foreignness into Mediterranean. Unlike the kites these waterbirds have expanded north without the hand of Man, unless it be climate change that's prompting them to tolerate our warming latitudes. And when I say north, I mean as far as Montrose Basin, where I recorded a little egret in 2003.

I had some trouble comparing this year's red kite sightings from this extremity of Ceredigion with last century's trip. I'd lumped all my south-west Wales records under Dyfed, which existed in the 1990s but no more. It's reverted to the three historic counties of Cardigan – sorry, Ceredigion – Carmarthen and Pembroke. Ancient counties that probably boasted a multitude of kites. And trees.

15th, 2008 The Yorkshire Wildlife Trust reserve at Potteric Carr is on the A6182 out of Doncaster, just over a large railway bridge. And trains form a backdrop to any trip here; indeed they even become part of the viewing from one of the hides. A visit is also a tale of routes, as befits the railway theme.

From the entrance I took the Green Route through alders, where siskins called but wouldn't show themselves. A boardwalk led me to a view over Decoy Marsh, which wasn't too busy, being mid-August. One reed warbler chattered in the *Phragmites*.

Across a freight line the Yellow Route branched left to the Childers Hide, which looked over Low Ellers Marsh. In between, abundant trains on the East Coast Main Line provided some amusement and made the whole experience a little surreal. The birding was still a touch quiet though – even further round the Yellow Route, which traversed excellent varied habitat.

It was only at the Red Route junction, with its view over the East Scrape, that waders began to make the day interesting. From nearby West Scrape Hide and along the northern edge of Huxter Well Marsh, I logged avocet, ringed plover, lapwing, common sandpiper and oystercatcher. Other waterbirds included little egret, a Mediterranean gull and unseasonal pochards. Sand martins and house martins zipped over the surface of the water while swifts patrolled overhead.

This put me at the Field Centre for lunch at Low Ellers Junction Café. Birdfeeders here and at Willow Pool Hide attracted the usual tits, finches and so on. I hoped for the alleged willow tits but only heard one marsh tit.

I was back on the Green Route and completed its loop round Loversall Pool and back to the entrance via a bridge over the Mother Drain. I love that name above all the others: it sounds so Yorkshire and a fitting end to a few hours in the south of the county.

16th, 2010 An adult spotted flycatcher was feeding two juveniles in the evening at Ham Wall, on the Somerset Levels just to the west of Glastonbury. It's a stonker of a bird in any case but, after torrential rain on the way down, I'd have settled for anything.

The day was clearing to promised sunny spells. A splendid male marsh harrier made good viewing as I plodded further in to the reserve, but only as far as the information board and screens. I could have gone on to look for the reported little bittern but didn't fancy the odds of finding it. Not compared with the chance of finding the real purpose of my trip at Priddy, or more accurately Stock Hill, some ten miles north on the Mendips.

This was nightjar. Although when I got to the pine plantation, the habitat didn't seem quite right and then the weather kicked in yet again. You win some, you lose some. But mainly you lose some.

18th-20th, 2011 I should know that when a weather forecast says showers, it translates to solid, incessant rain in Welsh. That's what I got from Rhayader, past the Elan Valley, through Devil's Bridge and all the way to Cardigan. It only eased there and on past Moylgrove. All the time I was thinking, "I'm only two or three hours from Bristol. I can always run back there."

But I pushed on to Fishguard. Did I see anything through all this precipitation? One red kite, just in Herefordshire, before I hit Knighton. Another somewhere through Powys and a wheatear as I entered Ceredigion. But above all, the streams were in full spate and silver filaments decorated the hillsides. That's the blessing of rain over mountain roads. Wouldn't swap it for a lottery win.

Even so I hoped for a clearer day after at the tip of Pembrokeshire.

It dawned. Wet.

No matter. My B&B landlady assured me that I could park right on Strumble Head, for that top seawatching site was my goal.

It was also windy there. Again no matter. The car made an acceptable hide and seabirds were close enough to shore – gannets, guillemots, one razorbill, kittiwakes and Manx shearwaters.

Manxes are so oceanic that any opportunity to watch them is a gift. Not even one in a hundred holidaymakers gazing out to sea knows the creatures exist so the birder has a whole extra dimension to enjoy. The shearwater colony at Skomer must be the closest to Bristol so the Pembrokeshire coast must also be the closest true seawatching experience. At 160 miles it is more than three hours distant – not exactly a day trip.

Then I somehow spent the rest of the day pootling in the county. I even broke from birding and dropped in to touristy St. David's.

So I was late afternoon driving into Pembroke itself for a reported scaup. Roadworks slowed me and on the spur of the moment I stopped a further night. I ticked the scaup – the only bird in the entire Castle Pond; it was almost eerie – ate a very hot Madras and was ready for another seabird city at Stack Rocks on the Castlemartin MoD range the following morn.

But only if the Army had played along, or rather didn't want to play with their toys. Red flags were flying and the barriers were down, just like old days at Barry Buddon in Angus. They could have at least blown something up for me as alternative entertainment. Never mind: I added yellowhammer to the Welsh list at the end of the Angle Peninsula and scuttled for home when heavy showers settled in.

19th, 2009 A song thrush has treated me with early morning visits to my temporary residence in Walkwood, Redditch. My mate, Dave, has feathered his nest here.

The usual fare is blackbirds, including one juvenile that pecks at nearly everything in sight – I guess it'll learn – so it's good to see the speckled member of the Turdidae family.

I associate the first inkling of massive population declines in birds with song thrush. Ten years ago it starred in that sort of headline until a dismaying cohort of other species joined it – skylark, yellow-hammer, marsh tit, grey partridge, bullfinch, corn bunting, *inter alia*, and now even house sparrow and starling.

Walkwood does hang on to a few marsh tits, bullfinches, sparrows and starlings, more so in the winter. In July the most obvious birds seem to be wood pigeons, the jumbo jet of the back lawn. As I type, a jay flies across, becoming more obvious daily in its quest for early acorns.

Having bred for the year, black-headed gulls are back on the school playing field and regularly lope over the house. A sparrowhawk – a female, it was so large (too much to hope for goshawk in these parts!) – flap-flap-glided past as I pulled in from shopping yesterday. And of course I see the usual tits, finches, corvids, dunnocks and robins. We don't need to worry about their populations.

Do we?

Andy Gibb

August

1st-5th, 2009 BirdLife International has identified 190 critically endangered species of bird. The
criteria for inclusion on the list are statistical and
hard to grasp. The best meaning I, as a poker
player, can give relates to gambling: don't bet on a
given species making it into the next decade. The
bet makes even more sense when nine out of ten
may already be extinct, have no known population
or have one that is still declining.

These critical 190 species constitute two percent
of the world's total for birds but Europe has only
two of them. One is barely European, being stuck
halfway across the Atlantic on the Azores. There the
endemic bullfinch numbers fewer than one thousand individuals and has pretty much nowhere else
to go.

The situation for the continent's next most vulnerable species is not so clear. It looks better, with
a population somewhere between five and thirty
thousand, although also declining rapidly. This bird
has much greater mobility, which along with its
habitat, causes the uncertain population estimate.
It is a seabird.

It is the Balearic shearwater (scientific name,
Puffinus mauretanicus, where   is the genus to
which the species belongs).

I had the privilege of joining SeaWatch SW for
five days to count these ocean wanderers past

Gwennap Head, just south of Land's End. In the previous two years, they had numbered about a dozen a day, so I could have hoped for around sixty on my watch.

Land's End. That's in Cornwall, right? Balearic, that sounds Mediterranean? Right again: those Spanish islands of Mallorca, Menorca, Ibiza and so on. What are the birds doing up here? Therein lies part of the problem. We're all aware there are no longer plenty of fish in the sea and this especially holds for the Mediterranean. The shearwater is foraging this far just to eat.

That's a hell of a distance to travel. It's not exactly nipping round to the nearest McDonald's. It is yet another amazing avian feat. A further wonder is how near we can push birds to their breaking point. Still, this particular pressure does at least give the Brits an opportunity to monitor the species locally – something we do well.

Myself, not so well in this case. I'd only ever seen one Balearic shearwater, somewhat streakily off a cruise through the Bay of Biscay. My view had been fleeting and far away, which is normal for seabirds. In these conditions, the species looks similar to its abundant British cousin in the same genus, the Manx shearwater.

Dark top half, light bottom half is the trademark colouring for the genus and Manx shearwater is the palest below with only brown edgings to its wings. The Balearic adds a few smudges fore and aft and under the armpits to give a more diffuse gradation from dark to light. It takes a trained eye to separate the two species.

Andy Gibb

Fortunately we had a couple in the head of John Swann, resident of Cornwall and veteran sea-watcher. Without him on the team I wouldn't have coped, being challenged enough by the prospect of getting up pre-dawn to spend twelve hours on an exposed Cornish headland. If the summer had been anything but the monsoon we'd started with, I might have relished the idea. As it was, I packed everything waterproof and windproof that I owned.

I took a couple of days to drive down from Redditch and so was able to preview the site the morning before my shift started. To log all the Balearics, which tend to pass closer to shore, the team had set up camp lower than the well-known watch-point at Porthgwarra. I scrambled down unforgiving Cornwall granite and through a defile we got to know as The Crack of Doom. Here I met John, and Russell Wynn, co-ordinator of the project.

Preliminaries over, I decamped to a B&B, which came as part of the deal, and then finished the day as a tourist: i.e. driving into and out of Land's End without paying the four quid to stop there; and failing to find anywhere at all to park in St. Ives. It was August.

One other observer was an integral part of the team while I was on Gwennap Head. Julie Hitchins, coincidentally also down from Redditch, was tallying dolphins, sharks and seals. The project's remit covered more than birds.

On the first hour of the first day, she and I counted nothing as we struggled to build a shelter from a couple of umbrellas and a few tent ropes. We'd both seen the intended construction the day before but without registering the exact configuration. We did finally settle down to surveying but didn't fare much better until John arrived.

There's a technique to this lark. There's also the proper equipment. John had a telescope with a wide-angle, 30x magnification eyepiece. My tiddly 25x zoom simply didn't trap enough light or give a decent field of view. I almost fared better with 10x42 binoculars.

The technique relied on knowing which general route the shearwaters would take. The Balearics would more or less accompany the Manxes, which we were also counting. These were flying on an east-to-west line about a kilometre out. In theory, scanning this route in the opposite direction should have caught most of the birds. In practice, the sea was pretty samey and offered little guidance to keeping a telescope on this line.

So, using it correctly was just as important. The answer was not to scan at all but to park the scope on a hotspot. Conveniently, the Runnelstone buoy sat about a mile offshore. Keeping this at the top of the view kept the line of flight in the bottom. It was less interesting and active than randomly swinging round the entire ocean, but more reliable.

The best seawatching spots in the country have such offshore markers. They also serve to guide other observers on to a sighting and to find what someone else (usually John) has seen. It was best to have my scope parked there when not attending to it.

After a while, I could make out some patterns in the target patch of sea, thanks to prevailing winds and underlying topography, like a reef in our case. They gave me more landmarks (seamarks?) on which to anchor and I could even scan between them.

In the course of this experimentation, I discovered a feature of my telescope, unsuspected for six years of ownership. I had chosen my Kowa because I found that its image wavered less than in my experience with other scopes. It still tended to wander and slide like a film trying to escape its screen, but it never disappeared completely.

Probably like all the scopes I had previously tried (belonging to other folk), a rubber rim surrounds the business end of the eyepiece. During my watch I must have absently rolled it back.

Blimey! Suddenly the image was solid and with a greater field of view. The rim was only the thickness of a pound coin but it had been sufficient to push my eye too far away.

Why does that matter?

This tendency to waver (technically *vignette*) depends on the position of the eye relative to the eyepiece lens. With luck one can hit its sweet spot by placing one's eyebrow on the eyepiece but faces and eyes vary greatly and no lens can cater to them all. The first resting place may not be ideal; indeed one may have to hold one's head away from the scope.

Most eyepieces allow some adjustment here, even if it's only rolling back the rubber rim. This is generally to accommodate spectacles but those with deep-set eyes find it helpful too. I do have deep-set eyes. Think Tom Cruise but without the rest of the looks (or the money). Opticals designed for the average face don't cut it for mine.

Inspired by this discovery, I tried looking through my binoculars *sans* glasses. Same result: instant increase in steadiness and clarity. It was an epiphany, spending five days learning how to use my binoculars and telescope. I'd never been in such a long relationship with them.

I also learned how to manage my expectations. When you see a figure like 3,024 Manx shearwaters in that period, it's tempting to think they were streaming past. Divide that number by the 3,600 minutes that five twelve-hour days represent and the stream diminishes to a trickle. Add the fact that shearwaters also tend to clump into groups of anything up to twenty and there's actually an awful lot of quietness out there.

We did have the constant company of gannets. Who could complain about that? The local shags and fulmars also went about their business. And the paperwork kept us busy. Every hour we monitored the state of the wind, clouds, sea and its glare.

To get an accurate reading for the wind meant a scramble up 50 metres to the cliff top. Twelve times a day for five days... strewth, I climbed the equivalent of three Munros. No wonder my packed breakfast disappeared faster each day.

The rhythm of scanning, recording and monitoring banished boredom and became the normal stuff of life. The rest of the world lost focus, turned unreal, irrelevant. Michael Jackson could have come back to life and we wouldn't have noticed.

* * * *

Andy Gibb

I was yet to settle in to this pace when the initial Balearic slipped past at 7:36 on that first Sunday morning. This set off another rhythm – ten that day; fifteen the next; and so on. We were counting beats, beats in the passage of sea life.

By eight o'clock the first of very few auks – a razorbill – had entered the records and my wind-climb produced a peregrine falcon. On the next hour I missed a passing great skua. Another pattern developed; it became a standing joke that the skuas would wait for me to start my ascent.

My compensation on that nine o'clock climb was ravens and, at ten, a wheatear. During the ensuing hour the Balearics peaked at four and I finally had a good enough view of one to add it honestly to my British list.

The morning ticked on. At twelve we broke for an hour, the period of maximum glare, then settled back to the rhythm. Before long, a sooty shearwater appeared, looking like a gigantic swift skimming the ocean. Another British first for me, we counted twelve in the entire five days.

A land-based distraction punctuated the afternoon: a pair of choughs probing the turf on cliffs behind us. Meanwhile on the non-avian front Julie was having a quiet time until a basking shark, close by. For me the day was indeed full of firsts.

And the weather held. We needed the umbrellas against twelve full hours of sunshine.

It was a different story on the Monday. The wind wrecked our construction and drove a spray up into our optics. Again, we needed the arrival of John who moved us farther up the cliff to get out of the weather's worst. Still, visibility was bad with frequent showers and often the only object of interest was a rock just below our viewpoint.

The tide was such that an occasional wave would wash over the rock, leaving on its surface a pool that gradually disgorged back to the sea through a couple of waterfalls. One of the flows was robust but the other would sometimes dry up before the next replenishing wave. I fell into a reverie of imagining the falls in Yosemite and how they operated on the same principle, except with snow as their reservoir. A mild panic even began to set in when, rarity of rarities, the main waterfall threatened to falter.

It never did. Whoosh, wave. Pool, falls. Vigorous waters, slowing. One fall, drying, sometimes dried. Whoosh, wave...

That's what the hours can do to you. Again, it was a different pace of life, relaxing into the pulses of the natural world. The day passed quickly. Before I knew it, I was back at the B&B and marking up the final totals for an email to Russell and BirdGuides. My job wasn't over until the paperwork was done and the paperwork showed 25 Balearics for the two days – bang on target thus far.

I didn't need the alarm to wake me on the third morning. The weather did that. It was foul. The road from the B&B to Porthgwarra was a river and there was no question of donning anything but waterproofs to trudge to the watchpoint. No question of struggling with umbrellas either. We moved location again to get out of the worst of the wind-driven rain. The showers did abate within a couple of hours and settled into a variety between thick fog and barely Runnelstone-disclosing mist.

Even so, we kept counting birds and conditions improved post-lunch to a beautiful afternoon, then a glorious evening. I got my first puffins of the watch and, in a grand gesture of profligacy, missed the week's only Cory's shearwater on my four o'clock ascent of the cliff.

We had company too with a few other hardy birders dropping in – as part of their holidays! It may have been fresh pairs of eyes but they always got on to birds quickly. And Limpy, the bent-legged herring gull, from previous years, came to watch us, watching them...

So we rolled on to day four, back down at our starting location in better weather. After lunch we even needed the umbrellas again in their parasol capacity. If there's an opposite term to stir crazy but with the same symptoms, I think we began to experience it. The Crack of Doom got its name. The local seals were becoming so familiar that we were now noticing "the beautiful female".

We began spinning fantasies around the pecking order of the Cornish fishermen out in the bay. They seemed not to bother with steering their boats and we imagined this was some sort of macho expression of how experienced they were. We hoped no Cornishman captained the odd oil tanker that glided by on the horizon.

That afternoon we became captivated by the odd massive wave that some strange combination of tide and wind brought in. We would not have noticed that before. We were becoming part of

Gwennap Head.

The fifth day dawned, and remained, breathless. The occasional mist formed, then burned off. The world was so still that, for the first time, we could hear the bell on the Runnelstone, tolling as though some watery church were calling the sea-faithful to their prayers. I resorted to fluttering a handkerchief to get any direction at all for the wind; I could no longer feel it on my skin.

The day's only major incident was purely personal when I discovered a scratch on my glasses from just placing them on the ground. That Cornish granite bristles with quartz, which only needs to brush glass to dig into it. Thank God, my binoculars didn't suffer the same fate.

So, what of the Balearic shearwaters? I had anticipated about 60 and we logged 51 over five days. It's too small a sample to say anything but, coupled with records over the entire season and compared from year to year, a picture will emerge. A picture into which I was glad to donate my pixel's worth.

And what if I had to get soaked and blown around on a Cornish headland to do it? The natural world needs as much help as we can give. Mind you, five days on the Azores, counting bullfinches... Hm, there's a drier-sounding idea.

17th, 2012 In this wettest of summers a rare dry, hot day allowed me an excursion to East Sussex. Arlington Reservoir was first in the hope of osprey or black terns. No result, so I tackled narrow lanes down the Cuckmere Valley and the wider A259 along to Beachy Head's western spur.

Here, Birling Gap is a National Trust honeypot but one minute away lies a less disturbed treasure – the Lookout. Even a baking afternoon couldn't deter migrants from using this hillside of scrub and wildflowers as a refuelling stop before their overnight flight across the Channel (English not Bristol, mind!)

My initial yomp up to the Belle Tout Lighthouse hadn't been promising with a few fly-past sand martins and two or three wheatears at the top. My return, on a more inland route, brought me my first whinchat of the year, which was probably also a migrant. Possible breeding birds were lesser and common whitethroats and a rather splendid female kestrel.

Chalkhill blue butterflies adorned the whole walk and I rounded off with a pint of Beachy Head Original at the Tiger Inn, East Dean. It's what summer should be made of.

Penultimate Weekend

The British Birdwatching Fair at Rutland Water coincides with the start of peak migration. Perhaps it is cunning planning by the organisers. They know we're champing at the bit and migrants are dropping in to the site's lagoons and scrapes. Common and curlew sandpipers, ruff and snipe join the earlier waders there – still in small numbers but the volume grows as September comes round and soon bodies of water throughout the whole country can boast dozens of these birds.

Family parties of tits and finches have been apparent for a while, buzzing, trilling and ticking through treetops. Warblers mix with them – chiffchaffs, blackcaps, whitethroats. Other passerines pass through – swallows, whinchats, house martins, spotted flycatchers, yellow wagtails, you name it. Following them are raptors, principally ospreys, hobbies and marsh harriers, and our prayers go with them ahead of their hazardous Mediterranean crossing.

The whole bird world is on the move again. Now is the time to check through those flocks for vagrants, individuals off their beaten track due either to bad luck or bad genes. Anything can blow in, and anything does.

The birdwatching community is back up and running, with a full wind in its sails.

21st, 2012 Dungeness, only just in Kent, was more than a two-hour drive from West Sussex thanks to the mess of Bexhill and Hastings on the way. This put me at the RSPB's shingle reserve in the afternoon doldrums and, apart from sand martins, a lone great white egret lurking at the back of Denge Marsh was the highlight.

That was until I added the ARC pit (even our natural world is subject to the creeping miasma of corporations) to the roster. Then the shorebirds took off. It started innocuously enough with lapwings but hit its stride with obvious migrant waders.

One juvenile little ringed plover was easy to tell beside several common ringed plovers. Dunlin and common sandpipers joined what may have been a local redshank. A marsh harrier, which isn't a wader, perched in trees backing the scrapes and one garganey represented the migratory ducks.

Meanwhile a couple of thousand more sand martins swooped and skimmed over the water. I've never seen such a cloud of the birds. A lady in the hide likened them to a swarm of gnats – but much less irritating.

This finale left me no time to check pools on the way back over the county border. However, it took me less than two hours to regain West Sussex despite a westering sun rendering the road ahead less than visible.

24th, 1999 My fondest sighting of an osprey was not in Scotland but on a trip back to Bristol, at the nearby Chew Valley Reservoir. It was my first true wild one, as opposed to those released at Rutland Water every year.

Bucketing rain had trapped me in a hide for an hour and I was considering making a run to the car when a rather large gull hove into view. On second look it was far too large a gull and the fingered wing tips gave me osprey before I had any other field marks.

The bird flapped and glided over the lake and that sight alone would have been worth it but I got several minutes of it before the predator wheeled and dived right in front of me. After it hit the water a flurry of wings and droplets obscured the action but the avian giant finally rose with a huge silver fish in its talons. The victim wriggled and thrashed but the bird didn't let go. It couldn't: its talons are designed to grip that fish no matter what. And no matter what can mean death. Trout in particular will dive and one strong enough will drag a weakened osprey under water.

This bird though continued to beat its way into the heavens. Then, in a remarkable manoeuvre it shuffled the fish into a forward facing direction. I prayed that it wouldn't let go before raptor and prey flew off over the hide.

I let my breath go and realised just how long I had been holding it, willing the osprey to keep its catch.

September

2nd, 2012 More than a year and a few cancellations after booking a Bristol Channel cruise on the Balmoral, there was fair enough weather to go. Not great weather: drizzle and cloud were constant companions and a strong headwind caused all manner of problems.

First of these was the actual boarding at Clevedon Pier, which is more exposed to the swell than most. Then we ran late all the way outbound and the English side of the paddle steamer Waverley – for that was the bonus vessel on this rescheduled trip – was too blowy for comfortable viewing. Some passengers were also clearly green about the gills.

Even I was a touch queasy until after settling with a fine breakfast for a modest £5.75 (including coffee!) I've survived New Zealand's notorious Foveaux Strait and Poole Harbour, which can be almost as choppy in a small boat. California's Monterey Bay pelagics too.

I was hoping that this 14 hour sailing would be an equivalent pelagic for the West Country. Sadly, the Bristol Channel is rather devoid of seabirds, especially in post-breeding September. The first hour across to Penarth – still in the Severn Estuary – was never going to deliver (unless in gales) and breakfast occupied most of it anyway.

Then I settled in on the Welsh side of the boat,

due to the aforementioned wind, and basically watched Barry, Aberthaw Power Station and the Glamorgan coast recede. Occasional forays to the Somerset board only disclosed the other Channel power station of Hinkley and the clear white parasols of Butlin's at Minehead.

A moment of excitement ensued as we drifted closer to Exmoor: a dark long-winged bird sped across our bows. It was too fast and direct-flying to do more than discount shearwater and any auk. Skua was possible but shag or cormorant was more likely.

Periodic breaks to warm up from the birding were spent overlooking the Waverley's three steam pistons pounding away and generating heat. This was oil-fired steam, it must be said, but still worth it for the smell and action. And warmth.

The birding improved as we crossed into Devon, halfway between Porlock and Lynmouth. Foreland Point is the most obvious sign of the new county and it also marks the start of proper cliffs. You'd like to think they attract breeding seabirds but only on the stretch beyond Lynmouth. This carries on to Ilfracombe, where Exmoor plummets into the sea, and forms the highest cliffs in England. The odd fulmar did begin to appear. A few gannets and shearwaters also glided over choppy waves.

The Waverley's approach to Ilfracombe allowed views of a rather odd town, draped over hills and outcrops. Odder still was the town crier greeting the boat as it docked. It all looked worth investigation but we had an island to visit, another 20 miles on across the top of Bideford (or Barnstaple) Bay.

Here the trip list gained a couple of guillemots but the numbers of any species still weren't high. One gannet joined us and glided albatross-like off the port side of the twin funnels. Lazy wing flaps were sufficient to keep pace with our ponderous progress. The bird kept it up for several minutes and allowed plenty of snaps from on deck before drifting back to the ocean.

As Lundy, wearing a garland of mist, finally hove into view, I added one shag. Several more then loafed round the island before we docked, a good hour late.

Once ashore, a peregrine was the first landbird tick. It perched, surveying the disembarking throng, then powered off. The captain gave us an extra 30 minutes on the island but it was still only enough to climb to the Marisco Tavern for a pint. Not that the mist allowed much sightseeing anyway.

I soaked up not one, but two, pints of Old Light at the Tavern. I was either that thirsty or it was that good (it is actually rebadged St Austell Tribute, so good enough). Other birds were restricted to robins, sparrows, crows, swallows and one black-bird. Then it was time to trudge back down and repaddle the 80 miles back to North Somerset.

The wind had subsided so we didn't even get a compensatory push to make up some time. The calmer conditions did allow an extended view of the spectacular Exmoor cliffs. I soaked these up too. Then darkness fell for the final three hours of the return to Clevedon Pier where one of the year's higher tides was racing out past its pilings.

That was a final worthwhile sight for a day that had provided a good share of experiences despite the weather. Or maybe because of the weather. That's the beauty of Britain.

5th, 2009 Lord knows what our robins, blackbirds, tits and finches think in spring when waves of hirundines, warblers and swifts sweep into the country from the Third World. Is there an avian equivalent of the BNP? And would the blackbirds be welcome in it anyway? It's not even as though party leader, Griffon (Vulture), is native.

OK, that's stretching the parallel too far. Upton Warren in Worcestershire today saw a stream of house martins, swallows and sand martins pushing their way back to Africa. Hobbies were following: one perched by the Moors Pool and darted out for dragonflies. Another laboured over the hide with some unidentifiable feathered victim in its talons.

Birds of prey, in general, appeared to be on the move. I saw several buzzards, one of which looked distinctly osprey-like with long, straight, fingered wings. The light was appalling and I was trying to see through trees, so I had to write it off as yet another that got away. I heard later that a marsh harrier had passed at about that time but I would never have put my bird down as that.

No rarities, then. But some damn fine birds all the same.

9th, 2010 Water levels at Chew Valley Lake were low enough that a strip of mud in front of the Stratford Hide hosted five species of wader. They were a common sandpiper, lapwings, greenshanks, a snipe and a curlew sandpiper – a small number of birds but a nice variety.

In fact a flock of curlew sandpipers had flown over Heron's Green on my way in. I was rather proud of identifying them. They were calling and I binned them (BirderSpeak for "got them in the binoculars") but could only say they were dunlin-like. Too clean for dunlin though, which are also less probable in the autumn. The flock's calls matched the description in my *Collins* – definitely trilling curlew sands.

Heron's Green Bay also held two little egrets and three whinchats, for my second 2010 sighting of the chats. This made it a great year for them – nay, the best: I'd never managed a year with two records.

Black terns as well. Did I mention them? Only the reason for going to Chew in the first place, half-a-dozen were easy to separate from black-headed gulls way out in the middle of the Lake.

Despite being sated, I headed for more to the Bernard King Hide on the eastern shore. A solitary ruff was meagre reward but another wader tick. I had thought pickings would be better but the extensive mud round the hide is probably unwork-able for waders. The lake is not tidal, so its margins bake hard. We need some bird-crazy inventor to come up with a tide machine for our reservoirs!

I wasn't done with shorebirds. BirdGuides had a wood sandpiper at Blagdon the previous evening, so I negotiated the country lanes to Butcombe Bay and there was the little beauty. There too were another dozen or so greenshanks and common sandpipers and, as a grand finale, one spotted flycatcher hawking in the shoreline willows.

We're definitely into full-on migration. Goodbye to the summer doldrums, although not to its weather, which was perfect – just like the birding.

13th, 2012 The Brecon Beacons can be bleak, not just in terms of landscape but also of birds (and the one frequently implies the other). I took an opportunity to check this against the more westerly reaches of the National Park. And to explore a couple of famous Welsh valleys.

The first, running up through Aberdare, was inconsequential so I was soon heading into the Park. At its border a red kite boded well for the day. My target spot, up the course of Avon Llia, was pretty enough for a Forestry Commission plantation, thanks to the stream itself. Surrounding moorland provided a third habitat and plenty of edges between all three.

There was even a sheep carcass on the track down to the car park; scavengers could be expected! Indeed, more red kites obliged. A walk up through pines was quiet though, apart from one squawking jay and a few coal tits, goldcrests and long-tailed tits.

I was aiming for Ystradfellte Reservoir but downed trees blocked my way and forced me back to the moors, where I only expected meadow pipits. They didn't disappoint but a search along the Llia for dippers also produced two grey wagtails – always a result.

Keeping tabs on the occasional kite brought me the star of the day. This bird flapped low to the horizon above me. Flapped low, glided, then wheeled to display a perfect and unmistakable white rump. A ringtail! A female or juvenile hen harrier. My first in nearly three years since one near Brancaster in Norfolk.

How often have I scanned moorland ridges for just such a sight? And how often succeeded? If I say millions and none, it would exaggerate but the ratio would still be right. A first for Wales, this harrier hunted for a matter of seconds before the ridge obscured it. I tried to tramp higher but that was it for the show – a few seconds for a 200-mile round trip. Priceless.

Some of those miles on the return journey took me down the Rhondda Valley over from Hirwaun. This is where the last deep-coal mine in the valleys supplied Aberthaw power station until 2008. Nearby evidence still abounds of open-cast and drift operations. Which all rather prepares one for the grey grandeur of the head of Rhondda Fawr. Scoured deep by the last Ice Age's glaciers, the cliffs wouldn't have been out of place on Frodo's journey into Mordor.

Of course it was all once so much mellower before the coal industry stripped away the trees for pit-props. And where did that lead us? In part to the nasty towns that snake the rest of the way down the Rhondda. Our industrial masters call that Progress.

15th, 1999 At Cotswold Water Park I have a fixed set of locations from which I choose: always pit 34 towards the western end, at the official nature reserve, to check a logbook in the hide; nearby 44 for wintering smew; generally pit 68 in the south for waders; and so on. A change was on the cards and it was pitiful to ignore vast areas of the Park.

So, I trotted off round the lakes across the road from pit 34. The most northerly and still being worked, number 79 allegedly held waders and sand martins. Not this time, although I was probably too late for the martins. The jaunt only provided the mini-excitement of hearing a tern.

However, I decided to revisit the nature reserve. Pay-dirt time. A couple of large birds circling a way off above the horizon could only have been buzzards. Through my binoculars something did not seem right with them. Yes, they were riding thermals. Yes, they were big. Yes, the tips of their wings were fingered. But where was the dihedral – on either of them? They were both holding their wings flat, which was odd.

Honey buzzards started entering my crazed thoughts but these birds were way too dark. I was ready to settle for being fooled by a couple of outsize crows – again. Then one of them spread its tail into that diamond formation and I had the answer. Ravens. The obvious answer really but it was so unexpected for that part of Gloucestershire.

Confirmation of their size materialised when a falcon started harassing them. The bulky shape made this no kestrel, but a peregrine, which is about the same size as a crow. The interloper was considerably smaller than the two corvids. In fact it looked rather foolhardy. I had never thought that about a peregrine.

A quick sortie south to pit 68 rounded off my visit and brought an early golden plover. It also produced my first sparrowhawk for the Park. It's absurd how unusual birds, like ravens, can appear on personal lists before these more common ones.

15th, 2012 Passage waders on scrapes off the British Steel Hide are the name of the game at WWT Llanelli. On Saturday a few hundred redshanks surrounded a number of black-tailed godwits, several greenshanks and a spotted redshank.

I was lucky (or persistent) enough to pick that out as it largely slept and was in almost complete winter plumage. Its flight feathers though did show some spangling in contrast to the common redshanks. Of course when its head did go up, the finer, slightly decurved bill was a clincher.

Swallows and house martins added to the autumn migration feel, as did whistling chiffchaffs – heard but not seen. A flock of wigeon was in, probably fresh from their boreal breeding marshes and pools. If fresh could describe a bird that's just been scrabbling to bring up chicks, then flown a thousand miles.

Reported on BirdGuides and a stray from southerly climes, a spoonbill also spent the entire time asleep. Even so it was distinctive enough from half a dozen little egrets round about.

23rd, 2012 I found choughs at RSPB South Stack, and Cemlyn Lagoon for complete overkill. Both sites are in north-west Wales on Anglesey, or technically Holy Island for the RSPB reserve, which is just past Holyhead. I was last there in June 1999 when the cliffs were full of breeding seabirds. Not so this year: they'd dispersed although a few gannets fished offshore and four razorbills bobbed around Cemlyn Bay.

That was a surprise. Also a surprise was a few turnstones for their first appearance on my Welsh list. And then of course the choughs. I'd made a special point of hunting them down at South Stack and had found three individuals. Then two at Cemlyn just fell into my lap.

So that was the choughs – the purpose of the trip. I may have also tracked an arctic skua until my attempt to zoom too far in lost it. I'm beginning to wonder about the eyepiece on my Kowa scope. I've already mentioned how ineffective it is as the image darkens, beyond 40x magnification as it happens. 30x wide-angle would be a better idea, especially for seawatching.

Anglesey is now an easy day trip from Snowdonia thanks to the A55. Back in 1999 it took a grind along the A5 and I elected to overnight in Holyhead, which was... ahem, dull. This year though I could base myself near Llanberis.

What amazes me is that in my time I've driven all four sides of the Snowdon massif and never been in. Now I realise it's the kind of area where I could happily die, being similar to the Highlands. It also recalls bits of the States. And seabirds are nearby. All a mere 130 miles from Bristol, although it takes a while to drive: the baleful motorway has not penetrated the fastness of Wales. Long may that continue!

26th, 2009 Doubtless there were more but an hour's stroll surprised me with twenty-one robins a-singing in a small Redditch coppice. They were second only to wood pigeons whose abundance was not a revelation.

Robins are more obvious at this time of year, being, along with wrens, about the only bird in full cry. It's all territorial now but even so the song sounds more melancholy than aggressive. There's the pitfall of applying human interpretation to a wild creature. Robins also depart from most birds' behaviour with the female being as adept a singer as the male. She tones it down somewhat in the spring when it is still the bloke's responsibility to advertise the goods.

Now, I'm guessing here because Google has failed me, that the sexes don't have much to do with each other from the autumn onwards. So, each individual bird maintains its own territory, which only breaks down once spring arrives.

I also watched three buzzards soaring. How pleasant to see them without fear of some gun-toting thug blasting them out of the sky. I felt almost shell-shocked from the slaughter I'd witnessed in Malta earlier in the month. I was thankful to be back in less murderous England.

October

3rd-4th, 1999 Had the preceding night been a shocker or what? The hotel car park had developed lakes and the sky did not bode much better as I drove out to Rattray Head, Aberdeenshire. Indeed a thorough drenching curtailed that little idea – and all I saw was one seal.

So, a mite damp, I retreated to the visitor centre at Loch of Strathbeg. A passage spotted redshank brightened the day and newly returned flocks of pink-footed geese kept it interesting. These centres frequently accommodate some guy who insists on calling out the birds he sees. This example pointed out two whooper swans, which I had passed over as mute swans. OK, OK, these people are a good idea!

My excuse for missing the whoopers is that they were asleep. Even so, they were a delight. They seem wilder than their mute cousins and therefore more impressive.

I motored through the old aerodrome to hides between it and the Loch, which gave me better views of a whooper flock. I also added scaup to my British list after plenty of waiting for it to float past tufted ducks to compare sizes, head shapes and so on. Again, the bird was asleep, so the job was problematic, but I was ultimately happy to tick it.

The day petered out after that. I was wearying of being blown about and rained on, so I settled for driving round ports on the north Aberdeenshire coast – including Pennan where Bill Forsyth shot *Local Hero*. I don't recall the weather in that film being as dreich as my experience.

Lossiemouth overnight was an unexpected treat. Quite apart from being a pleasant town with a real ale bar, it then supplied purple sandpipers, bar-tailed godwits, ringed plovers, turnstones, red-breasted mergansers, a stonechat, a rock pipit, gannets off-shore, and a shag in the harbour.

The last of these proved difficult to identify. I am not used to seeing shags, so I had to convince myself that its forehead really was steep and it had very little white on the chin. More persuasive was the way that it looked smaller than nearby herring gulls. I had never actually compared herring gulls with cormorants (never had to!) but made a mental note to do so next time.

As I got back to the car three small streaky finches perched on a low wall. They were not linnets and they could not have been any of the arboreal finches: the sea-front at Lossiemouth is not over-endowed with trees. I thought they had a pinkish wash to the breast but could not remember if that made them twite or not. I waited for them to fly to get the colour of their rumps but that was inconclusive when it happened. My field guide suggested they should have had a buff wash and was particular about the colour of their bills. I had not noticed that. Still, it was really just a choice between twite and corn bunting and the latter was so unlikely that Sherlock Holmes himself would have plumped for the former.

I stopped in Elgin, having failed to find a way in

to Lossie Forest. The plan was to head down the Spey valley. However, in my idleness I found a bookshop and browsed a map of the area. This showed a large forest at Culbin. I was getting too much storm-blown coast and wanted some trees for a change. So, I drove out towards Findhorn and Culbin Sands.

Not good for one small bird.

I am good at keeping an eye out for all wildlife on the road. I have actively avoided several animals in my time. This day though the small shape on the road did not register as a bird. It was so motionless. Only as I was almost on top of it did it try to fly away with disastrous consequences. I could see it, still, on the road behind me before another car passed over it. I'd had a brief glimpse of white before it flashed under my bonnet and thought that it was probably the rump of a bullfinch. The middle of the road is not a likely place for one of these but perhaps that accounted for how slow it was to react to my car.

That took the edge off the day. It was all so unnecessary. Why did I have to be driving along that road right then? Only to tick some more names off an imaginary list. Pointless, pointless, pointless.

And irrelevant to add that I finally caught up with crested tit at Culbin Forest.

6th, 2014 The rains have returned. After the driest September on record an October wind howled and a deluge beat down overnight in Newquay. September had also reverted to being a couple of degrees above the latter part of last century (itself warmer than the preceding century).

Business as usual will be back – floods, recriminations, promises, retractions – and all the while a blackout on the "global warming" phrase.

A breezy morning walk round Newquay's golf course worked off my cooked breakfast. A lone skylark broke into untimely song. Jackdaws chacked but no choughs materialised. A handful commutes the North Cornwall coast between Perranporth and Padstow. They could have been anywhere on that stretch: much of the terrain looks suitable – rough grassland and cliffs.

The odd beach and crashing surf along the way also draws in pleasure-seekers and Newquay itself is devoted to them. My B&B though was discreet and not likely to attract that rowdier element.

Close offshore from Towan Head gannets glided and plunged. Did the storm bring them in? Among dozens of birds were a sole juvenile and two first summer individuals.

Odd. Gannet numbers have been on the up for years although they're still amber listed: their population is highly concentrated round few sites, so they're vulnerable to the sort of ecocide that industrial civilisation loves to dish up.

Why then just one juvenile? The correct question should be: why was there one at all? Fledglings immediately leave their breeding colony and head south, far south, even to the equator. Adults tend to disperse more locally, if North Africa could be called local.

It was time for me to disperse, but a much lesser distance, to Padstow for the first time in forty years – maybe. University friends and I had been in the area back in 1976 and it seems unlikely that we wouldn't have visited the town. Either way, I bet it wasn't as precious as it is now; you even have to pay 20p for a piss. I gave the money-sink short shrift and explored the Camel Estuary.

At least I tried. A more detailed map of the area would have helped as I drove down several dead ends. Indeed I did miss a car park halfway along. One road was noteworthy though, again harking back forty years. Above the old railway line from Wadebridge still sits Tregunna House; that's where we students had stayed. The trackbed is now the Camel Trail and it must be the best method for working the estuary; there is also more accessible parking at the Wadebridge end.

Nothing dismayed, I moved on to Crowdy Reservoir at the northern edge of Bodmin Moor. There were no beasts to speak of but a late wheatear and a flock of ringed plovers – no little ringed among 'em: they're long gone by now. A lone dunlin was strange but the species will turn up anywhere.

The reservoir is hard by Davidstow Airfield, a rather minor WW2 (and Falklands) station. 'Tis an eerie landscape. Crumbling runways crisscross a vast expanse, parts of which have been planted with conifers. This conjures images of hen harriers and goshawks coexisting in the winter but the area turns up more waders than raptors.

That was about it for the day apart from one final twist. I thought Okehampton, by Dartmoor this time, would be good for an early dinner break. Not so. But a glance as I crossed a stream rewarded me with a speeding dipper. It's funny how that species turns up in the least birdy places.

16th, 1999 Focalpoint's bird news had reported a late Temminck's stint on its way, probably from northern Scandinavia, to Africa via Titchfield Haven. Hampshire Ornithological Society's mailing list confirmed it. I sped straight to the reserve in the morning.

I need not have hurried. The place didn't open until nine-thirty, so I had time to admire a flock of bearded tits flying and feeding at the reed tops. The lighting was superb and they were easy to identify even in mid-air. A few Cetti's warblers also called but typically didn't show.

Come the half-hour, I wended into the Meon-shore Hide, almost full already. The comments as I entered confirmed that something of interest was out there. I was cool. I even removed my rucksack before sitting down. Hey, I'm English! Then I looked in the general direction of everyone else's scrutiny.

There was only one bird in the entire half of the pool and it was easy to identify as a stint. Again, the light helped and showed the colour of its legs – pale compared with little stint's black. So, separating the two was not difficult. I was interested in how differently it fed from the little stints I had seen – very leisurely, very deliberate.

I watched for thirty minutes, during which a green sandpiper and a water rail also showed. The stint continued to perform and even latecomers were not disappointed.

It was a beautiful day, so I covered the rest of the reserve. I didn't go far from the hide before hearing a little owl although I couldn't locate it. The place seemed to be brimming with jays, sparrowhawks and green woodpeckers. And later, I finally glimpsed a Cetti's.

18th, 2010 A couple of Bewick's swans on the Rushy Pen trounced my earliest autumn sighting for them by nearly a month. They're so early that you have to go back to 2003 to find Slimbridge's own equally early record for them. Now some are predicting a harsh winter because of this but I was in Auchmithie, then Dundee, over 2003/4 and I don't recall any severity. I think the swans are early because they're early.

[Indeed late November and December 2010 did produce an extended bitterly cold and snowy spell (see below) but that was it. The succeeding January and February were mild – so mild that I was registering butterflies. Three and a half years since have continued the trend of 2°C above late 20th-century temperatures, themselves higher than the previous century.]

A nice number of pintails complemented the Bewick's and a grey wagtail finally made it on to my species list for the site. As I left, my first fieldfares of the winter flew over the In Focus hide.

26th, 2011 Not exactly in Bournemouth, Fisherman's Walk Zig Zag runs down the cliff at Boscombe. That is in Bournemouth Unitary Authority and hence technically no longer in Dorset, so in my listing world Hampshire can reclaim the site.

Yes, this is important. It's one of only two counties where I've seen Dartford warbler. In any case, this bird was in urban, built-up Bournemouth. Quite a contrast to my previous rural sightings at Acres Down in the New Forest and Slapton Ley, Devon – both in 1999.

Twelve years before! Imagine my disbelief then as I nipped out between heavy showers to admire the clouds and light at the Overcliff Nature Reserve and this *Sylvia* warbler popped up in front of my nose. The bird gave no cause for binoculars: its profile was so distinctive. Even so I grabbed great close-ups as it foraged through the cliff-edge gorse. It was like winning the lottery without entering.

More predictable later in the day was dipping on a red-breasted goose at Stanpit Marsh, Christchurch. The reserve was hard to find and then its entrance path was under water. Boy, it has rained. I know: I'd driven through hail, lightning and road-rivers at Yeovil the day before.

29th-30th, 2009 The Lancashire leg of my Glasgow trip started at WWT Martin Mere, which late October means whooper swans. They didn't let me down, especially as feeding time approached. They became my 654th species in twelve months, starting with Singapore the preceding November.

As well as the swans, thousands of pink-footed geese filled the Mere's pools and fields with their incessant chatter. The sight of yet more flying in to join them is one of the British autumn spectacles. I soaked it up in a couple of hours and various hides, towards the end of which came the real highlight.

A highlight more because I had forgotten it was likely and so hadn't anticipated it. Right at the other end of the size spectrum, a small colony of tree sparrows persists on the reserve. A species I had last seen in Singapore in fact, where they are city birds. I'd dipped on them in the intervening year, even at the BirdFair, where they normally haunt the feeders.

Coincidentally the evening before driving up to Martin Mere, BirdFair's founder and Rutland Water good guy, Tim Appleton, had been speaking. I caught him as part of the excellent programme put on by the Kidderminster branch of the West Midlands Bird Club. He'd made the point that the feeders at Rutland operate year round. None of this mothballing in spring and restarting in autumn.

The interim summer denial of service does seem so mean to me, especially as breeding birds need their own nosh while they have the added pressure of bringing up baby. Thanks to Rutland's generosity, its tree sparrows fledge three broods a year, compared with the national average of three or four chicks.

Back to Martin Mere, where human chicks were evident because it was half-term. Many a grizzled birder (guilty me too!) would complain about the resultant noise and whatnot. However, it's the kids who, in a couple of decades, may be deciding the fate of everything from tree sparrows to whooper swans. If those decades leave humanity with any oil, water, food, etc. but that's another thread for another book...

I spent the night at Morecambe, which has plentiful, cheap accommodation and one superb Chinese restaurant. And then there's the Bay. When the tide's right, the shore drips with waders – the next morning: oystercatchers, redshanks, a lone dunlin, black-tailed godwits and one bar-tailed godwit; and turnstones.

I love turnstones. They can be so confiding as though they don't realise you're there. They are handsome creatures too with plumage whose colour fits their formal name of ruddy turnstone. In winter the ruddiness disappears but like most waders both sexes moult into colour for breeding. Unlike most waders turnstones will feed on almost anything, even human corpses. We're not so sacrosanct after all.

There's been some tinkering with waders' taxonomy under the IOU but they still basically split into plovers and sandpipers. The turnstone sits between the two large sandpiper genera of *Tringa* and *Calidris*, represented by greenshank and curlew sandpiper respectively. In Europe the common sandpiper also occupies its own odd little niche here. Odd little niche – maybe that contributes to the turnstone's appeal too. It feels like a quirky bird.

Ten miles further up the road was a single pinging bearded tit at Leighton Moss. The RSPB reserve there also held four snow geese in the Allen and Morecambe section. Here the Morecambe is named for our beloved Eric of TV comedy fame.

He was a birdwatcher. Him, Bill Oddie, Rory McGrath, Bill Bailey, Steve Martin... there must be something about birding and being funny. Also novelists: Ian Fleming, Margaret Atwood, Jonathan Franzen, Agatha Christie. But most impressive is the clutch of presidents: Teddy Roosevelt being the most obvious, but Jimmy Carter too and, apparently, Fidel Castro.

Great company!

November

1st, 2010 Proving that a pint is good for you, three nights ago a tawny owl called once as I pulled into the car park of the White Hart. This is the village pub for Weston-in-Gordano, halfway between Portishead and Clevedon.

The single note, rising at the end, was enough to separate the call from that of little owl. The smaller bird tends to keep on an even keel or downslur except for one of its sequences, which includes a longer preamble.

I had to wade through more than two minutes of tawny owl noises on my Roché CD to be sure: the species is almost as profligate as great tit in its repertoire. But I was happy enough with the identification and I'd finally logged all five British owls in the one year.

It was a surprise that the commonest of our owls was the last, and so late – the end of October. Fully half my tawny owl records have occurred in the autumn. Since this is a bird ten times more often heard than seen, one might jump to the conclusion that this is their favourite season for hooting.

So, November is upon us. Chaffinches are back; they don't frequent Port Marine during the summer, my last record there being in March. A handsome male sparrowhawk also perched openly this morning by Portbury Rhyne, and a huge flock of long-tailed tits kept me hoping for something more exotic among them. Not that I was disappointed: those pint-sized fluffy aerial tadpoles are a delight any time.

1st-2nd, 2011 At first it looked like a buzzard, scattering the wigeon, teal and pintail at RSPB Pulborough. Then it banked to show a creamy white head and its true identity: a marsh harrier. It must be said that a juvenile, and maybe a female, has similar colouring to a dark morph buzzard so one does need the head for a positive identification. That or a good few moments to observe the harrier's quartering behaviour.

My bird landed straightaway and disappeared behind low scrub so that luxury wasn't on offer. I wondered if it had found a meal when it failed to reappear.

This excitement followed stunning views of a female peregrine falcon. Both these birds of prey were new for my Pulborough list as were a couple of snipe and a calling crossbill. I may also have seen a female crossbill at the very apex of a conifer but at the distance she was hard to separate from a greenfinch.

To continue the raptor theme the next day, my trip back to the West Country brought me two red kites. Not up the A34 nor along the M4 as one might suppose but through the heart of Hampshire, near Old Winchester Hill. That's my third record in the area so they are drifting south.

And they made a raptorlicious couple of days.

6th, 2017 A frosty old start to the day, typical of the east coast. It took a few minutes to scrape the ice off the motor and a only few more to be stopped by an accident. I could have driven past but, as second on the scene, I felt that the girl dealing with it probably needed some support.

A kid had come off his wee moped and though he didn't seem too hurt, you can't take chances, so I hung around for the medics to arrive. The lass who was calling them was a kid too – just passed her test. This all occupied half an hour but at least I wasn't having to get to work or college.

No, I was aiming for south of Felixstowe. From Ipswich this should be simple enough but like all English towns Ipswich is plagued with traffic. On more than one score I'm praying for oil to double back to over $100 a barrel: some sanity may then prevail. So long did my journey take that two coffee stops extended it further and it was afternoon before I finally stepped out on Landguard Nature Reserve.

Actually I also tried to detour via Trimley Marshes but that's a long old walk in and I doubled back once the racket from Felixstowe Docks got too intrusive. This did put me in place for a marvellous view of something brownish plummeting from the treeline into a margin of long grass. My first thought was kestrel but it didn't reappear and I began to wonder if it could have been a large dead leaf.

Then the creature lifted and flapped away and in the bins I could see a body in its left talon. It's not often you witness a kill and I don't think I've ever seen a kestrel in the act.

These little falcons are legion in Suffolk. Some

stretches of road provided five or six of them. Farther west you're lucky to notch a couple in a day. This may not be unconnected with the paucity of buzzards in the east. Once red kites had taken over on my journey from Bristol through the Chilterns to when I later headed home through Bishop's Stortford, the buteos were quite absent.

Landguard was about as exciting as Trimley. The sea produced four scoters, four Brent geese and a couple of Mediterranean gulls. No other bird was of note but I could still see the odd passerine straggling in off the North Sea, so the potential was clearly there.

Fish and chips beckoned and Felixstowe seemed a good bet for that. A stroll of the front delivered zilch apart from the observation that the flood gates close permanently during the winter. The sea is coming for us! This of course is the year that Houston, Florida and the Caribbean experienced that assault big time.

Anyway you'll be relieved to hear that Winkles, to the north at Felixstowe Ferry, was open for haddock and chips.

12th, 2003 I bagged a lifer – in among the housing of Craigie Drive, Dundee of all places – a flock of waxwings (Bohemian of course) trilled its way round the berry-bearing trees there. Down from previous days' counts of more than a hundred, there were still dozens of these irruptive visitors from Scandinavia.

I recently also added fieldfare and grey plover to my Angus list but dipped on snow goose and Slavonian grebe at Loch of Lintrathen over the weekend. The Scottish days are getting truly short. Perhaps owls will be easier to come by now?

12th, 2012 Hard by what should logically be called the West Hide at RSPB Minsmere in Suffolk, a bearded tit pinged in the reeds. I thought that would be it but the bird – strangely just one – kept calling from the pathside vegetation and I followed until I hit paydirt. A mere five feet from me the restless shape paused long enough for a sighting of warm brown tones and the moustachial comma that gives the species its name.

The bearded part of its name, that is. The bird isn't a tit though. It's near, but in a separate family – its own family moreover. The clan has no close relatives. What makes it an oddity is a phylogenetic (i.e. something to do with DNA) relationship with larks, which follow the Paridae in the IOU's taxonomy. *Panurus biarmicus* acts the missing link.

Anyhow, I hadn't seen bearded tit for two years since Leighton Moss. Another species new this year was yellow-legged gull; three stood out among streaky-headed herring gulls. This is the best distinction between the species when they're standing in water, as gulls tend to. It's then easy to persuade oneself that the inkling of leg on view is yellow.

Other early winter denizens of Minsmere were: one marsh harrier wisely not braving the killing fields of Malta; burgeoning flocks of shoveler, gadwall and wigeon; offshore goldeneyes, scoters, Brent geese and red-breasted mergansers; and not much in the wader department – a few lapwings and black-tailed godwits.

The next day's route back to the Midlands took

in Fen Drayton and Grafham Water – both in my near-virgin county of Cambridgeshire. The RSPB reserve at Fen Drayton produced the "usual suspects", including a very dark buzzard. One could have mistaken the raptor for that honorary raptor – a raven. Paths around the lakes were sodden after the wettest several months I can remember and I was glad of a tarmac road as an alternative route back to the car park.

Grafham was quiet soon after lunch and, worse, entirely without victuals in the café. A pair of bullfinches alone were of interest for a downbeat ending to a couple of exploratory days away.

13th, 2009 No Wilson's phalarope at Slimbridge for the assembled faithful this morning. Plenty of dunlin and golden plover from the Zeiss Hide and even a couple of black-tailed godwits, discernible through the murk. Enough to keep me occupied for an hour before the gloom turned to the forecast rain and I scurried back to the Visitor Centre.

Happily, Bewick's swans were back to complete our *Cygnus* species for the year: mute swans are ever present and I had already caught up with whooper at Martin Mere. It's good to have the entire family gathered in time for Christmas.

Bewick's, our smallest swan, is only on BirdLife International's Amber List. Only? That makes it a relative success story with the broad meaning that the species is of some concern. More specifically, Britain hosts a large proportion of the wintering population but split between few sites. The birds are vulnerable to a disaster at one of them.

For instance, it'll only take a meltdown at Berkeley nuclear power station to do for the Slimbridge population; but, hey, who'll be worrying about a bunch of feathers in that event? (It's OK: I know Berkeley's been decommissioned; but you get the general idea?)

Later, a distant party of white-fronted geese from the Holden Tower took me to my 660th species in the last twelve months. I hasten to add that this count is for the entire world thanks to my own wintering activities – to whit, travelling Australia, New Zealand and California at the turn of the year. Escaping one British winter. Two were about to arrive that would be worth escaping.

17th-19th, 2014 All the way along the M18, M62 and beyond to Kilnsea, recent rain had left standing water in many fields. Night fell; being north and east of Bristol, East Yorkshire loses its sun early. I had to search for my B&B in darkness and succeeded just as a sharp shower coincided with my dash between car and front door.

"You timed that well," said the landlady.

I'd heard that phrase before, in better circumstances.

Our introductory chat veered to the reserve on Spurn Point. "You'll need to check the tides," she said. "Where the road's been washed away, the waters meet at high tide."

"Since when?" I didn't remember this from June 2010.

"December."

Of course, a tidal surge had hit the east coast – potentially worse than 1953 but mitigated by flood defences and advance warning systems since. Even so, it did enough damage and Spurn was one of its casualties. I was curious to see how that would play out next day.

Well, my stroll from the new end of the road followed the route that 4WDs take to the lifeboat. On the face of it, sand had piled over the original road. It was only when I tramped the few yards to the edge of the North Sea that the truth emerged. There, the road was crumpled into its component sections.

That stopped me: a great deal of energy went into this destruction. Rising sea levels will clearly not be a gentle process of water lapping over the coast; there will also be violence and damage. And winners too: no more tourist vehicles at Spurn Point must be a Good Thing. Seven years later the road hasn't been rebuilt, joining the growing roster of evidence for collapse and how it proceeds – little by little, bit by bit.

The seawatching hide though was still drivable and I spent a couple of hours in the company of Yorkshire stalwarts on a pleasant November day. The action at sea was quiet, especially compared with previous reports of little auks, pomarine skuas and the like. Red-throated divers and common scoters did make an appearance to send my year list up to my best ever at 205. So I was happy.

Happier still to log number 206 up the coast at Hornsea Mere, thanks to a Slavonian grebe. By then it was only 2.30 but the light was fading and thoughts turned to the night's accommodation. Bridlington beckoned – the more so because it would give me a second shot at seawatching from

Flamborough Head the next morning.

This provided guillemots, razorbills, shags and fulmars but was otherwise no more successful and I turned west again and back to longer afternoons.

18th, 2009 Shore larks at Holme and a purple sandpiper at Sheringham – both in Norfolk – boosted my round-the-world twelve months to 662. Dave (remember him from July?) and I battled gales to steady our scopes on the larks and the winds hadn't much diminished an hour later at RSPB Titchwell farther along the coast.

The reserve was sheltered enough for passerines to flock round feeders near the Visitor Centre, and through a copse that enclosed it. But once trees gave way to reeds and the path continued along a raised bank... well, we were pleased to scurry for the Island Hide.

This was the only operational viewing point for the lagoons and scrapes. Parrinder, farther on, would reopen a year later with its wall strengthened against rising sea levels. Yes, the coast was also retreating in Norfolk. In 2012 I would witness the same engineering for the same reason at Stert in Somerset. Global warming is not a myth. It's a real fact with real costs: the Parrinder wall saved Titchwell from the December 2013 storm; nearby Snettisham was not so lucky, being almost wiped out. And we already know what happened at Spurn.

Leaving the Island Hide, Dave and I carried on to the end of the path and the encroaching sea. The beach held its quota of littoral waders, and Brent geese, goldeneyes and red-breasted mergansers bobbed on the more distant stormy swell. They at least may benefit from the surfeit of water in the world.

21st, 2014 A change of name for RSPB Burton Mere, from Inner Marsh Farm to something more descriptive of where it is, which should have helped after ten years of failing to find the place.

It wasn't as simple this time either and I again found myself hunting round Neston, where my road map incorrectly places the RSPB symbol. This lower half of the Wirral Peninsula in Cheshire is becoming rather familiar. Even so, it took the local postie to set me right and send me back to a very discreet sign pointing down Puddington Lane. Hey, how about calling the reserve Puddington Marsh? I like that name a lot better.

Anyway, I entered a visitor centre full of people looking at... a merlin. Nice, but not what I'd driven an hour from Warrington to see, and I didn't get on it all the same. A cattle egret though was showing a ways away in a field of cattle; that bird at least is well named. Despite the distance, it was unmistak-able with its hunched posture and truncated bill.

That gave me species number 207 for the year and I knew that 208 was around: a ring-tail hen harrier had been reported on and off. Sitting and waiting is not my *modus operandi*, especially with an entirely new site to explore, so I took off, with a weather eye on the reedbeds: harriers like those.

It was the weather weather as put paid to that plan, for it started drizzling and five minutes later I was back in the visitor centre.

"The harrier just flew into the reeds," announced one of the RSPB guides.

Oh, boy. How long would she stay there before showing again? Experience told me an hour or two. Happily, the rain didn't amount to much and I checked out the Marsh Covert Hide and a busy feeder along the Reed and Fen Trail. In all, my visit yielded 37 species in a couple of hours and yes, one of them was the harrier. She started quartering the reedbeds about 90 minutes after her initial appearance – about on schedule.

That made a productive morning and just as well because my route onwards along the North Wales coast took me through the heavy downpour that had been forecast. Mind you, it was clear by Bangor and the northwest corner of Wales, which must be in some sort of rain shadow. Either that or I've led a charmed existence there.

A few showers were the order of the evening and following morning. Being on a social visit, the birding was on a back-burner but I still managed to log great northern diver and rock pipit at Caernarfon. Sometimes one doesn't have to hunt too much.

27th, 2013 To acclimatise back to chillier conditions compared with Tenerife, I also treated the day after my flight as holiday. Isn't there always so much to do on one's return anyway? Even so I was clear by early afternoon although, in contrast to almost tropical abrupt sunsets at 6.30, an evening gloom was already spreading.

Reports of a black-throated diver at Chew Valley Lake had scrolled through Twitter for a week and I'd been willing the bird to stay. Now was my chance and I was on the 20-minute journey from south Bristol to Woodford Lodge. Here, a lone photographer stood. My pulse quickened.

"Oh yes, it has been close," he said. "I've lost track of it but it should still be around."

I scanned and was waylaid by several great crested grebes. In winter they're similar to the diver in colour scheme and structure, but sharper and more slender in detail. Those that caught the edge of my vision in mid-dive needed waiting for a reappearance to check these finer points. Nearby goldeneyes also compelled a longer linger.

A kingfisher zipped by for only my second Chew record and finally, not too far distant, I caught a trailing white flank patch that suggested black-throated diver. It plunged but, with the bird's location pinpointed, I could savour successive views that confirmed its grey head, scaled flight feathers and more substantial bill than red-throated, which is about the only confusion species.

Six years had passed since my previous sighting. Of course that was in Scotland, at Backwater Reservoir in Angus. Another six occurrences squeezed into 2003/4, which was my heyday for birding the Scottish wilderness.

The day at Chew Valley finished with the winter's first goosander, flying into the dusk. Then three days later my list gained two more species with a pair of pink-footed geese and a dozen starlings. The farthest south I'd ever seen the geese was Slimbridge so they were breaking new ground.

The starlings though were an oddity. I'd noticed their omission three years before. I'd thought they were an oversight of recording but, having taken so long then to appear, they must be scarce at the

Lake. Other species missing in 2010 were redshank and mistle thrush, which had turned up in the interim; however, greenfinch still eluded me in 2013.

A comparison with my county lists revealed the absence of many summer passerines too, probably because Chew is an autumn and winter trap. Spring and summer are more about the coast and hills in Somerset.

December

1st, 2010 Portbury Wharf was dripping with birds this morning, largely flocks of redwings and fieldfares. The ponds were iced up so the only ducks were flying but I still netted teal, wigeon and gadwall. A lone stonechat patrolled scrub near the Port Marine development.

Lapwings ventured on to the fields and snipe flew regularly. Sightings of just one of those signify a red-letter month; today the total was three. Skylarks and reed buntings were equally obvious; it was like the whole bird world was on show.

There was even a mystery individual. It flew over my head on the salt marsh. It could have been a plover with its small pigeon build and shape but it didn't fit grey plover: brown upperparts, no black axillaries and no wing bar. Golden? Unlikely all on its own and it dropped onto the tidal mud, so even less likely. A bird doesn't often flummox me and this went down as unidentified.

That was the best of the day. As I returned, the sun, which had kept it tolerably warm, retired behind thin clouds. Now it's raining.

4th, 2016 The bachelor finches had arrived in force. Dozens occupied most niches on the feeders at RSPB Lake Vrynwy in North Wales. A few females, one greenfinch, several tits and a nuthatch also managed a clawhold but male chaffinches dominated the scene. And they weren't at all fazed by the presence of humans about a foot away behind the windows of the Coed y Capel Hide. It was spellbinding. Of course I was greedy and hoped for a brambling among them but the year had already delivered those.

A crisp, sunny day was drawing to a close and the low light allowed easy spotting of three red kites overhead and two far distant ravens. A buzzard and a sparrowhawk completed the raptor line-up. But what of the lake itself?

Rather like Scottish lochs it was devoid of birds. There's something about these highland bodies of water that doesn't support much life. Earlier in the day Llyn Brenig, where new-fangled Conwy nicks into the resurrected Denbighshire, could only muster four mallards. Again, passerines rescued the site with siskins on the visitor centre feeders, and crossbills commuting between pine plantations round the reservoir perimeter.

Plenty of moorland on my circuit of the Berwyn Mountains was also lifeless, though permitting of some stunning views. Oddly my journey up through the roads of the Welsh Marches the day before had yielded more avian interest. I even flushed a snipe from the tarmac at one bend. But what was oddest was one blackbird.

At least I assume it was a blackbird because as it flew up, it was definitely black and the right size and shape. However, a white stripe ran across the middle of its tail feathers – a white stripe punctuated by a black gap so that the bird looked like it had the Morse Code for M imprinted on it. Dash-

dash in white.

No bird in the world fits that description so my blackbird must have had a pretty strange moult going on. Or I've described a species new to science.

10th, 1999 Birds on farmland? You should already know my thoughts about that but they weren't so well formulated back at the end of last century. There was an ongoing population crash, which began to suggest that farms would be the last place to look for birds. Yet the BTO's Winter Farmland Survey was taking me to a couple near Crawley, Hampshire for a day's counting. Oh well, it would keep me fit.

A cheeky chaffinch got the ball rolling before I was even out of the car. Across a field pheasants and red-legged partridges foraged. No surprise there: shooting interests abound in that neck of the woods.

I had to calculate a route covering a one-kilometre square with all its complications of field entrances in the wrong places and unmapped barbed wire. A slow scan with the binoculars for these hazards revealed a handful of fieldfares. No, about a dozen camouflaged against bare soil. No, there was another dozen. The dozens added up to a round hundred, which also hid a couple of starlings and a skylark.

My plod round the most apparently efficient course disclosed little further apart from a singing linnet. Singing? In December? Oh, and grey partridges. They took a moment to identify, lacking an obvious black horseshoe on their underparts; this made them probably all females.

Andy Gibb

So, then it was on to the square kilometre for farm two, the source of most of the game birds. I could have left the car where it was but not before returning to it for a cup of tea and my friendly chaffinch. As a diversion I got it in the bins and was puzzled by the colour of its breast. The same colour showed in its wings, and dark splodges patterned its back. No chaffinch then but a brambling.

My second route started with a tiny patch of scrub by a dilapidated barn. Less cultivated than the surroundings, it still just produced more red-legged partridges. Only when I headed away down a large field did the action get going again.

A buzzard floated over and circled. Meadow pipits and skylarks called and flew briefly. An electric chirp told of a hidden yellowhammer. Small birds commuted between a scrap of weeds and the bordering hedge. Most were chaffinches but some were more bramblings, some yellowhammers, and a few... reed buntings. I didn't expect them.

With all the comings and goings it was impossible to judge numbers. I had to guess, over the course of ten minutes, at about a hundred. This ate into my time budget enough that I would not cover the entire square.

I hurried on past a chiffchaff in the same hedgerow. This was only the second year I'd recorded the species' wintering behaviour. Three kestrels loafed by a barn. Another hundred fieldfares. Or the earlier flock? Who could tell?

A final treat was a bullfinch when I got back to the car. That and the earlier brambling were more typical behaviour: some of the best birds appear in the car park!

15th-16th, 2014 England has a second, mini-Lake District, or more accurately a Mere District. In northwest Shropshire its largest lake is at Ellesmere. During winter this has a 10,000-strong gull roost. Those with an idle few hours try to find species other than black-headed, common, herring or lesser black-backed.

I didn't have this amount of time at the end of an afternoon sampling the lesser bodies of water. They hadn't yielded much; in fact the best was a red-legged partridge flying up from the road! I hoped the ensuing morning session would be more exciting.

In the interim I did get to walk past the Mere in starlight. Gracious! There's one beneficial side-effect of a large expanse of water – diminished light pollution. Orion and Gemini were just rising; Taurus was riding high; and the Plough and Cassiopeia swung round the Pole Star. That's the limit of my heavenly navigation skills these days.

My morning half-circuit of the Mere (you can't go all the way round) supplied more goosanders, a few nuthatches, one treecreeper and... er, that's about it. No lesser spotted woodpecker, although my information there is some twenty years out of date and, given the species' massive decline, doubtless over-optimistic. No wintering smew either, so I then drove a good way in to Cheshire, where a female, aka redhead, had been sighted at Newchurch Common.

This doesn't sound very watery but looks to be a series of pits that have filled in. Situated in an angle between the A49 and A54, it's also not easy to find, probably because the Warrington anglers have sole possession of it and no intention of sharing. A spiralling route, blocked by dead ends, put me about as close as I'd get and then I tramped in some half a mile. About twenty minutes were needed to scan the lake until I finally latched onto the little beauty, tucked away in one wee corner.

20th, 2010 I kid you not. With Portishead Marina nearly iced over, one Mediterranean gull huddled with black-headed and herring gulls and – a patch tick – many common gulls. Indeed today has been a bonanza for patch ticks.

It started with a snipe careering over the house as I stepped out into the snow. I had notions of the bird being a woodcock but an obvious darker head and breast made snipe the only candidate. The third newbie, over Portbury Rhyne which was also turning to ice, was one fieldfare among redwings and blackbirds.

The snow has otherwise been keeping bird news sparse lately and almost trapped me in Kidderminster at the weekend. The stuff is following me around.

It's also cancelled tonight's owl prowl at the other end of Portishead so my town list for the year may have stalled at 106. Unless this weather brings in something truly exotic.

And that's always possible.

22nd, 2004 An unlikely direction for a day's birding was up the London-bound M4, and beyond the end of it. Crossing the Chiswick Flyover and down to ground level, where could I be going? The exit to Hammersmith put me on the A306 round-about, where the sign to Barnes prompted a quick switch of lanes over Hammersmith Bridge.

Yes, WWT Barnes Elms – five miles from the capital's centre and bang under Heathrow's landing path. Yet the site transcends all this to be the most wonderful metropolis wildlife experience in the world. We're so lucky.

But first I needed to recuperate at the Water's Edge café. It was on the route to the most interesting part of the site. The path beyond, to the Peacock Tower, also allowed drops in to Dulverton and WWF hides – all overlooking various lagoons.

A screech. A flash of bright green through upper foliage. A scan with the bins put me on to one of my target species but not one that should be expected. Rose-ringed (or ring-necked) parakeets have escaped captivity and set up a colony, growing across west and central London. They should be in a different continent, a different hemisphere even, but, in keeping with the city's cosmopolitan flavour, they added a touch of exotic colour to a drab winter's day.

Other, more usual, species, boosted by wintering ducks, soon had my visit's tally over thirty. A different route back from the Tower, past the Wader Scrape Hide, brought it up to 36 with snipe and pintail.

That was the wild side of the reserve and it easily soaked up the morning. Post-lunch I headed up through World Wetlands – aka the wildfowl collection – to a section with different views of the lagoons. No more species but I was happy with the final count for a December day.

22nd, 2011 Carrying on from Bournemouth, I had a beautiful saunter round Hengistbury Head, where ravens boosted my Hampshire list. I've said this before: Bournemouth itself and everywhere east is still in Hampshire for me. So, Slavonian grebes in Christchurch Harbour also counted, and Brent geese there were new for the year.

Then, best for last, an entirely hopeful punt south of Shatterford in the New Forest for a reported great grey shrike. A couple of birders coming the other way late in the afternoon hadn't seen it after a thorough search so I was anticipating at best another decent stroll. Not a bit of it: a white blob at the top of a birch stood out in the gloaming for a splendid shrike and I was able to get close enough for cracking views.

That's probably it for the year – 183 species – not as good as many years recently but then I haven't been gadding off down under or living in Scotland. The numbers don't count really. The birds do. And always will.

25th, 2013 A fine time to go birding is Christmas Day. Away from the excess and hypocrisy to celebrate in my own way. And where better than the Somerset Levels? Especially as they'd flooded, and evidence of that surrounded me on the way from Glastonbury to Meare. Indeed the road itself had clearly been inundated.

Wet weather from the beginning of the year was repeating itself. Was this a pattern for the future?

Only my later readers will know.

Shapwick Heath was the first stop, where three great white egrets were in attendance. On to Noah's Lake, which was higher than I'd ever seen but somewhat bereft of waterfowl. One kingfisher compensated.

There was then time to catch only my second visit to Westhay Moor. In the gloaming, snipe flew in convoy as skeins of starlings arrowed in and swirled a while above the reeds. A sprinkling of pied wagtails also formed a roost, as did a cacophony of jackdaws and rooks in the distance. A single tawny owl whitted and was silent. A fourth great white egret laboured away and closed proceedings for the day.

Twenty-four hours earlier little owls had entertained me at Portbury Wharf. There I'd also had to christen my wellies and wade in, to the reward of hearing my first water rail for the reserve. Earlier still a kittiwake off Battery Point had flown on to my Portishead list. The nearby boating lake had spilled over its margins so all around were signs of the deluge and high winds.

These continued into the New Year, to the extent that a spring tide up the Avon flooded parts of Bristol. This bodes well for seabirds and waterfowl. Not so well for humans and other creatures.

Summer Migrants

These are likely early and late dates for visiting birds. Note that some, marked by *, also winter in small numbers so the dates may refer to residents; blackcap and chiffchaff winter in such good numbers that I try to pick up the first and last singing birds.

Species	Early Date	Late Date
singing Blackcap*	27-Feb	27-Jul
White Wagtail	9-Mar	8-Oct
singing Chiffchaff*	11-Mar	25-Oct
Wheatear	11-Mar	8-Oct
Little Ringed Plover	11-Mar	29-Sep
Puffin	16-Mar	29-Oct
Sand Martin	16-Mar	12-Oct
Ring Ouzel	21-Mar	9-Oct
Willow Warbler	25-Mar	4-Oct
Sandwich Tern	25-Mar	12-Oct
Swallow	26-Mar	31-Oct
House Martin	27-Mar	15-Oct
Garganey*	29-Mar	21-Sep
Osprey	30-Mar	1-Oct
Green Sandpiper*	31-Mar	28-Oct
Sedge Warbler	2-Apr	6-Sep
Yellow Wagtail	2-Apr	15-Sep
Marsh Harrier*	3-Apr	23-Oct
Common Sandpiper*	3-Apr	27-Sep
Redstart	6-Apr	21-Sep

The British Birding Year

Whitethroat	8-Apr	17-Sep
Reed Warbler	8-Apr	15-Sep
Common Tern	9-Apr	28-Sep
Lesser Whitethroat	12-Apr	29-Aug
Hobby	12-Apr	6-Oct
Garden Warbler	13-Apr	21-Aug
Cuckoo	15-Apr	9-Sep
Wood Warbler	15-Apr	15-Jun
Tree Pipit	15-Apr	19-Aug
Grasshopper Warbler	15-Apr	13-Jul
Arctic Tern	17-Apr	11-Oct
Nightingale	17-Apr	24-Jun
Pied Flycatcher	19-Apr	27-Jun
Whimbrel	21-Apr	1-Nov
Little Tern	21-Apr	11-Jul
Wood Sandpiper	25-Apr	21-Sep
Swift	26-Apr	13-Sep
Spotted Flycatcher	2-May	19-Sep
Great Skua	9-May	19-Sep
Arctic Skua	10-May	13-Sep
Temminck's Stint	16-May	23-Oct
Turtle Dove	16-May	28-Jul
Nightjar	31-May	16-Jul
Curlew Sandpiper	5-Jun	6-Oct

Autumn Migrants

Here again, wintering birds tend to confuse the picture and it's hard to distinguish late passage individuals from them. In the case of green sandpiper, migrants can also be so early back that they almost overlap with late spring birds.

Species	Early Date	Late Date
Green Sandpiper	21-Jun	29-Oct
Common Sandpiper	1-Jul	17-Oct
Curlew Sandpiper	2-Aug	6-Oct
Wood Sandpiper	2-Aug	21-Sep
Black Tern	6-Aug	2-Oct
Garganey	10-Aug	21-Sep
Pintail	18-Aug	11-May
Little Stint	26-Aug	11-Oct
Pink-footed Goose	7-Sep	5-May
Velvet Scoter	15-Sep	26-Mar
Whooper Swan	21-Sep	16-Apr
Barnacle Goose (wild)	25-Sep	6-May
Brent Goose	27-Sep	21-Apr
Purple Sandpiper	1-Oct	29-Apr
Redwing	5-Oct	16-Apr
Brambling	10-Oct	21-Apr
Fieldfare	11-Oct	3-Apr
Snow Bunting	14-Oct	6-Mar
Short-eared Owl	17-Oct	25-Mar
Bewick's Swan	18-Oct	28-Feb

The British Birding Year

Long-tailed Duck	19-Oct	13-May
Great Northern Diver	28-Oct	21-Apr
White-fronted Goose	29-Oct	25-Mar
Red-necked Grebe	2-Nov	15-Feb
Waxwing	12-Nov	15-Jan
Bean Goose	14-Nov	15-Jan
Water Pipit	29-Nov	6-Mar
Smew	10-Dec	11-Apr

Glossary

BirdLife International – a global alliance of conservation organisations working together for the world's birds

BTO – British Trust for Ornithology

Class – such as *Aves*, birds. Other similar classes are mammals, reptiles, amphibians and so on to make up the vertebrates. See Order.

Endemic – occurring only in a restricted geographic location, which can be as big as a country.

Family – a subdivision of an Order of organisms. A family such as the chats and flycatchers, aka Muscicapidae, contains species, like stonechat and robin, grouped by Genus. See also Order.

Genus – a group of Species with some sufficiently big difference from another Genus, such as *Erithacus* within the Muscicapidae family. Its name forms the first word of the Scientific Name, e.g. *Erithacus rubecula* for robin. See Family.

IBA – Important Bird Area as identified by BirdLife International.

IOU – International Ornithologists' Union, source of the taxonomy used throughout this book, despite what any other field guide may say.

LBJ – little brown job, always a passerine that

skulks.

Life List – an enumeration of all the species of bird seen throughout the world, and so...

Lifer – a new species for the Life List.

Lores – the facial feathers between the eye and the bill.

Mega – a bird that's beyond rare for a given location.

Order – a subdivision of a Class (such as *Aves*, birds) of organisms. See also Family.

Passerine - an Order of birds that includes more than half the species in the world. Basically they're the littler birds towards the end of most field guides – the birds that perch and sing and visit feeding stations. In other words, not your waterfowl, seabirds, raptors, gamebirds, pigeons or woodpeckers.

Patch List – like a Life List but restricted to a locale, which could be as small as a garden or as big as a city. In any case the area should be covered on foot.

Pill – a not-strictly West Country word for creek or inlet.

RSPB – Royal Society for the Protection of Birds

Scientific Name – under the system devised by Linnaeus the label for a Species, such as *Podiceps cristatus*, the first word being the name of its Genus. Always italicised, it's the *lingua franca* for taxonomists around the world in preference to localised common names.

Tick – an addition to any list, such as Life List, Patch List, county list.

Twitch – a flocking of birdwatchers to a rare or mega bird.

WWT – Wetlands & Wildfowl Trust

My British List

This is a checklist of British birds but not as you will have seen it before. It's ordered from my experience of the species most likely encountered downwards and should give novices some idea of what birds they are probably looking at. Note that exclusion from the list does not mean the bird cannot be seen in the UK – just that it's less likely.

Wren	*Troglodytes troglodytes*
Robin	*Erithacus rubecula*
Blackbird	*Turdus merula*
Crow	*Corvus corone*
Chaffinch	*Fringilla coelebs*
Wood Pigeon	*Columba palumbus*
Blue Tit	*Cyanistes caeruleus*
Dunnock	*Prunella modularis*
Black-headed Gull	*Chroicocephalus ridibundus*
Mallard	*Anas platyrhynchos*
Jackdaw	*Coloeus monedula*
Great Tit	*Parus major*
Greenfinch	*Carduelis chloris*
Pied Wagtail	*Motacilla alba yarrellii*
Goldfinch	*Carduelis carduelis*
Herring Gull	*Larus argentatus*
Starling	*Sturnus vulgaris*
Magpie	*Pica pica*
Cormorant	*Phalacrocorax carbo*
House Sparrow	*Passer domesticus*

Andy Gibb

Coal Tit	*Periparus ater*
Song Thrush	*Turdus philomelos*
Buzzard	*Buteo buteo*
Mute Swan	*Cygnus olor*
Pheasant	*Phasianus colchicus*
Grey Heron	*Ardea cinerea*
Oystercatcher	*Haematopus ostralegus*
Swallow	*Hirundo rustica rustica*
Long-tailed Tit	*Aegithalos caudatus*
Lesser Black-backed Gull	*Larus fuscus graellsii*
Rook	*Corvus frugilegus*
Goldcrest	*Regulus regulus*
Kestrel	*Falco tinnunculus*
Feral Pigeon	*Columba livia*
Meadow Pipit	*Anthus pratensis*
Coot	*Fulica atra*
Lapwing	*Vanellus vanellus*
Moorhen	*Gallinula chloropus*
Curlew	*Numenius arquata*
Tufted Duck	*Aythya fuligula*
Sky Lark	*Alauda arvensis*
Chiffchaff	*Phylloscopus collybita*
House Martin	*Delichon urbicum*
Common Gull	*Larus canus canus*
Collared Dove	*Streptopelia decaocto*
Redshank	*Tringa totanus*
Mistle Thrush	*Turdus viscivorus*
Great Black-backed Gull	*Larus marinus*
Linnet	*Carduelis cannabina*
Canada Goose	*Branta canadensis*
Sparrowhawk	*Accipiter nisus*
Shelduck	*Tadorna tadorna*
Teal	*Anas crecca crecca*

The British Birding Year

Willow Warbler	*Phylloscopus trochilus*
Great Spotted Woodpecker	*Dendrocopos major*
Swift	*Apus apus*
Bullfinch	*Pyrrhula pyrrhula*
Blackcap	*Sylvia atricapilla*
Wigeon	*Anas penelope*
Great Crested Grebe	*Podiceps cristatus*
Reed Bunting	*Emberiza schoeniclus*
Jay	*Garrulus glandarius*
Greylag Goose	*Anser anser*
Little Grebe	*Tachybaptus ruficollis*
Green Woodpecker	*Picus viridis*
Stock Dove	*Columba oenas*
Gadwall	*Anas strepera strepera*
Pochard	*Aythya ferina*
Siskin	*Carduelis spinus*
Eider	*Somateria mollissima*
Dunlin	*Calidris alpina*
Shoveler	*Anas clypeata*
Yellowhammer	*Emberiza citrinella*
Goldeneye	*Bucephala clangula*
Redwing	*Turdus iliacus*
Nuthatch	*Sitta europaea*
Shag	*Phalacrocorax aristotelis*
Grey Wagtail	*Motacilla cinerea*
Turnstone	*Arenaria interpres*
Whitethroat	*Sylvia communis*
Sand Martin	*Riparia riparia*
Gannet	*Morus bassanus*
Treecreeper	*Certhia familiaris*
Red-breasted Merganser	*Mergus serrator*
Ringed Plover	*Charadrius hiaticula*
Black-tailed Godwit	*Limosa limosa*

Andy Gibb

Common Tern	*Sterna hirundo*
Stonechat	*Saxicola rubicola*
Fieldfare	*Turdus pilaris*
Rock Pipit	*Anthus petrosus*
Sedge Warbler	*Acrocephalus schoenobaenus*
Peregrine Falcon	*Falco peregrinus*
Fulmar	*Fulmarus glacialis*
Snipe	*Gallinago gallinago gallinago*
Guillemot	*Uria aalge*
Kittiwake	*Rissa tridactyla*
Marsh Tit	*Poecile palustris*
Pink-footed Goose	*Anser brachyrhynchus*
Razorbill	*Alca torda*
Ruddy Duck	*Oxyura jamaicensis*
Cetti's Warbler	*Cettia cetti*
Little Egret	*Egretta garzetta*
Reed Warbler	*Acrocephalus scirpaceus*
Bar-tailed Godwit	*Limosa lapponica*
Common Sandpiper	*Actitis hypoleucos*
Sandwich Tern	*Thalasseus sandvicensis*
Raven	*Corvus corax*
Pintail	*Anas acuta*
Wheatear	*Oenanthe oenanthe*
Goosander	*Mergus merganser merganser*
Green Sandpiper	*Tringa ochropus*
Hooded Crow	*Corvus cornix*
Golden Plover	*Pluvialis apricaria*
Greenshank	*Tringa nebularia*
Red-throated Diver	*Gavia stellata*
Water Rail	*Rallus aquaticus*
Whooper Swan	*Cygnus cygnus*
Grey Plover	*Pluvialis squatarola*
Little Ringed Plover	*Charadrius dubius*

The British Birding Year

Cuckoo	*Cuculus canorus*
Red-legged Partridge	*Alectoris rufa*
Knot	*Calidris canutus*
Ruff	*Philomachus pugnax*
Kingfisher	*Alcedo atthis*
Long-tailed Duck	*Clangula hyemalis*
Tree Sparrow	*Passer montanus*
Scoter	*Melanitta nigra*
Sanderling	*Calidris alba*
Lesser Redpoll	*Carduelis flammea cabaret*
Tawny Owl	*Strix aluco*
Spotted Flycatcher	*Muscicapa striata*
Dipper	*Cinclus cinclus*
Velvet Scoter	*Melanitta fusca*
Puffin	*Fratercula arctica*
Avocet	*Recurvirostra avosetta*
Red Grouse	*Lagopus lagopus scotica*
Hobby	*Falco subbuteo*
Arctic Tern	*Sterna paradisaea*
Garden Warbler	*Sylvia borin*
Grey Partridge	*Perdix perdix*
Osprey	*Pandion haliaetus*
Red Kite	*Milvus milvus*
Little Owl	*Athene noctua*
Brent Goose	*Branta bernicla*
Lesser Whitethroat	*Sylvia curruca*
Red-crested Pochard	*Netta rufina*
Purple Sandpiper	*Calidris maritima*
Slavonian Grebe	*Podiceps auritus*
Willow Tit	*Poecile montanus*
Brambling	*Fringilla montifringilla*
Crossbill	*Loxia curvirostra*
Tree Pipit	*Anthus trivialis*

Andy Gibb

Barnacle Goose	*Branta leucopsis*
Marsh Harrier	*Circus aeruginosus*
Merlin	*Falco columbarius*
Egyptian Goose	*Alopochen aegyptiaca*
Smew	*Mergellus albellus*
Black Guillemot	*Cepphus grylle*
Bewick's Swan	*Cygnus columbianus bewickii*
Redstart	*Phoenicurus phoenicurus*
Curlew Sandpiper	*Calidris ferruginea*
White-fronted Goose	*Anser albifrons albifrons*
Manx Shearwater	*Puffinus puffinus*
Spotted Redshank	*Tringa erythropus*
Little Gull	*Hydrocoloeus minutus*
Great Skua	*Stercorarius skua*
Twite	*Carduelis flavirostris*
Corn Bunting	*Emberiza calandra*
Bittern	*Botaurus stellaris*
Hen Harrier	*Circus cyaneus cyaneus*
Wood Sandpiper	*Tringa glareola*
Arctic Skua	*Stercorarius parasiticus*
Bearded Tit	*Panurus biarmicus*
Wood Warbler	*Phylloscopus sibilatrix*
White Wagtail	*Motacilla alba*
Yellow Wagtail	*Motacilla flava flavissima*
Mandarin Duck	*Aix galericulata*
Scaup	*Aythya marila*
Black-throated Diver	*Gavia arctica*
Great Northern Diver	*Gavia immer*
Waxwing	*Bombycilla garrulus*
Whimbrel	*Numenius phaeopus*
Little Stint	*Calidris minuta*
Rose-ringed Parakeet	*Psittacula krameri*
Snow Bunting	*Plectrophenax nivalis*

The British Birding Year

Goshawk	*Accipiter gentilis*
Woodcock	*Scolopax rusticola*
Mediterranean Gull	*Ichthyaetus melanocephalus*
Barn Owl	*Tyto alba*
Ring Ouzel	*Turdus torquatus*
Whinchat	*Saxicola rubetra*
Black Grouse	*Lyrurus tetrix*
Spoonbill	*Platalea leucorodia*
Pied Flycatcher	*Ficedula hypoleuca*
Scottish Crossbill	*Loxia scotica*
Ptarmigan	*Lagopus muta*
Snow Goose	*Chen caerulescens*
Pale-bellied Brent Goose	*Branta bernicla hrota*
Dark-bellied Brent Goose	*Branta bernicla bernicla*
Ruddy Shelduck	*Tadorna ferruginea*
Garganey	*Anas querquedula*
Red-necked Grebe	*Podiceps grisegena*
Black-necked Grebe	*Podiceps nigricollis*
Glossy Ibis	*Plegadis falcinellus*
Temminck's Stint	*Calidris temminckii*
Red Phalarope	*Phalaropus fulicarius*
Iceland Gull	*Larus glaucoides*
Black Tern	*Chlidonias niger*
Turtle Dove	*Streptopelia turtur*
Lesser Spotted Woodpecker	*Dendrocopos minor*
Chough	*Pyrrhocorax pyrrhocorax*
Crested Tit	*Lophophanes cristatus*
Yellow-browed Warbler	*Phylloscopus inornatus*
Nightingale	*Luscinia megarhynchos*
Black Redstart	*Phoenicurus ochruros*
American Wigeon	*Anas americana*
Surf Scoter	*Melanitta perspicillata*
Great White Egret	*Ardea alba alba*

Andy Gibb

Golden Eagle	*Aquila chrysaetos*
Stone-curlew	*Burhinus oedicnemus*
Pectoral Sandpiper	*Calidris melanotos*
Short-eared Owl	*Asio flammeus*
Hawfinch	*Coccothraustes coccothraustes*
Ring-necked Duck	*Aythya collaris*
Lesser Scaup	*Aythya affinis*
Barrow's Goldeneye	*Bucephala islandica*
Storm Petrel	*Hydrobates pelagicus*
White-tailed Eagle	*Haliaeetus albicilla*
Rough-legged Buzzard	*Buteo lagopus*
Red-footed Falcon	*Falco vespertinus*
Common Crane	*Grus grus*
Jack Snipe	*Lymnocryptes minimus*
Baird's Sandpiper	*Calidris bairdii*
Glaucous Gull	*Larus hyperboreus*
Yellow-legged Gull	*Larus michahellis*
Little Tern	*Sternula albifrons*
White-winged Tern	*Chlidonias leucopterus*
Nightjar	*Caprimulgus europaeus*
Great Grey Shrike	*Lanius excubitor*
Shore Lark	*Eremophila alpestris*
Iberian Chiffchaff	*Phylloscopus ibericus*
Dartford Warbler	*Sylvia undata*
Water Pipit	*Anthus spinoletta*
Capercaillie	*Tetrao urogallus*
Quail	*Coturnix coturnix*
Taiga Bean Goose	*Anser fabalis*
Tundra Bean Goose	*Anser serrirostris*
Ferruginous Duck	*Aythya nyroca*
King Eider	*Somateria spectabilis*
Balearic Shearwater	*Puffinus mauretanicus*
Sooty Shearwater	*Puffinus griseus*

The British Birding Year

Cattle Egret	*Bubulcus ibis*
Spotted Crake	*Porzana porzana*
Red-necked Phalarope	*Phalaropus lobatus*
Sabine's Gull	*Xema sabini*
Roseate Tern	*Sterna dougallii*
Long-eared Owl	*Asio otus*
Bee-eater	*Merops apiaster*
Red-backed Shrike	*Lanius collurio*
Woodlark	*Lullula arborea*
Red-rumped Swallow	*Cecropis daurica*
Pallas's Leaf Warbler	*Phylloscopus proregulus*
Marsh Warbler	*Acrocephalus palustris*
Grasshopper Warbler	*Locustella naevia*
Firecrest	*Regulus ignicapilla*
Lapland Bunting	*Calcarius lapponicus*
Rosy Starling	*Pastor roseus*
Common Rosefinch	*Carpodacus erythrinus*
Bluethroat	*Luscinia svecica*

Andy Gibb